M000268211

CHASING AUTOMATION

CHASING
AUTOMATION

THE POLITICS OF TECHNOLOGY AND JOBS FROM THE ROARING TWENTIES TO THE GREAT SOCIETY

JERRY PROUT

NORTHERN ILLINOIS UNIVERSITY PRESS
AN IMPRINT OF CORNELL UNIVERSITY PRESS
Ithaca and London

Copyright © 2022 by Cornell University

All rights reserved. Except for brief quotations in a review, this book, or parts thereof, must not be reproduced in any form without permission in writing from the publisher. For information, address Cornell University Press, Sage House, 512 East State Street, Ithaca, New York 14850. Visit our website at cornellpress.cornell.edu.

First published 2022 by Cornell University Press

Library of Congress Cataloging-in-Publication Data

Names: Prout, Jerry, author.
Title: Chasing automation : the politics of technology and jobs from the roaring twenties to the Great Society / Jerry Prout.
Description: Ithaca [New York] : Northern Illinois University Press, an imprint of Cornell University Press, 2022. | Includes bibliographical references and index.
Identifiers: LCCN 2021061459 (print) | LCCN 2021061460 (ebook) | ISBN 9781501763991 (hardcover) | ISBN 9781501764004 (epub) | ISBN 9781501764011 (pdf)
Subjects: LCSH: Technological unemployment—Political aspects—United States. | Labor supply—Effect of technological innovations on—Political aspects—United States. | Technological unemployment—United States—History—20th century. | Labor supply—Effect of technological innovations on—United States—History—20th century.
Classification: LCC HD6331.2.U5 P76 2022 (print) | LCC HD6331.2.U5 (ebook) | DDC 331.12—dc23/eng/20211228
LC record available at https://lccn.loc.gov/2021061459
LC ebook record available at https://lccn.loc.gov/2021061460

Amid the rows of modest, white headstones on the gently rolling hills that make up Arlington National Cemetery stands one that reads, "Lloyd A. Prochnow, USNR, World War II; December 1, 1923, to December 16, 1966." After his military service, Prochnow began work as an economist at the Department of Labor. His section at the Bureau of Labor Statistics would later work on the report issued by President Lyndon Johnson's National Commission on Technology, Automation, and Economic Progress. This book is dedicated to the service of Lloyd A. Prochnow and countless others who each day devote their lives to serving the public good and who are central to the story in the pages that follow.

CONTENTS

ACKNOWLEDGMENTS

This book began with a journey to the Lyndon Baines Johnson Library in Austin, Texas, in 2017. There I was fortunate to have the guiding hand of the archivist Allen Fisher, who had carefully assembled materials germane to my research on the National Commission on Technology, Automation, and Economic Progress.

As the manuscript progressed, I was fortunate to be able to rely on the insights of historians at George Mason University, including my mentors Jack Censer and Michael O'Malley, and the critical eye of Sam Lebovic. I also valued the encouragement and support of colleagues at Marquette University, including the accomplished political scientists Lowell Barrington, Julia Azari, and Paul Nolette.

Returning to Northern Illinois University Press, which published my first book, *Coxey's Crusade for Jobs*, senior acquisitions editor Amy Farranto welcomed me back and skillfully guided me through the peer review process.

In preparing this book for publication, my appreciation for my Cornell University Press production editor Jennifer Savran Kelly and my copyeditor Eric Levy grew daily. I came to admire their assiduousness and dedication to detail and style, so essential to the method of writing history, and I thank them for their significant contribution to improving this book.

CHASING AUTOMATION

Introduction
Seeing Forward

How can the future become a matter of public opinion?
—Bertrand de Jouvenel, as quoted in Bell,
"Government by Commission"

Robert Wagner was nine years old when his family moved from a rural village in Germany to dilapidated tenement housing in New York City. The youngest of seven children, "Bobby" took odd jobs hawking newspapers and delivering groceries, excelled in the city's public schools, and graduated from City College of New York in 1898 and from New York Law School in 1900. As he reflected later, "I came through it, yes. But that was luck, luck, luck! Think of the others."[1]

It was "the others" that Wagner cared most about in a career of public service that began in the New York State Assembly in 1905 and that catapulted him six years later to the role of majority leader of the New York State Senate. As the dedicated cochair of the investigation of the Triangle Factory fire, an advocate for worker rights in the state senate, and the author of pro-labor opinions as a New York Supreme Court justice, he had a reputation for progressivism in Albany that preceded his twenty-two-year US senatorial career in Washington. The first three bills he introduced as a US senator in 1928 were consistent with both this reputation and his upbringing: they required the government to develop a better method for counting the unemployed, allow the jobless in one state to learn about jobs in another, and provide for escrowing monies during good times to fund public works jobs in bad times. Throughout his senatorial career he championed laws that cemented organized labor's bargaining rights, protected the common laborer's wages and hours, and compensated the unemployed. As one observer summed up his progressive record, "The New Deal owed as much to Robert Wagner as to Franklin Roosevelt."[2]

In 1909, just as Wagner was beginning to carve out his reputation in the New York legislature, a fourteen-year-old child prodigy named Norbert Wiener received a BA from Tufts University. At fifteen he began graduate work at Cornell, completing his PhD at Harvard when he was nineteen. A mathematical genius, he studied under Bertrand Russell at Cambridge, then secured a post at MIT, where he began research that would radically transform the workplace three decades later. Fascinated by the randomness of the swirling waters in the Charles River, he began writing a mathematical language that further explained Brownian motion.

Though Wiener would not confer the name "cybernetics" on his new statistical mechanics until 1947, his theory began to take shape during his tenure at MIT in the 1920s. As his biographers note, it was with a series of obscure publications that Wiener provided a foundation for those sciences that spawned an automated age. Yet his breakthrough discovery of a new binary language was hardly noticed amid the din of an earlier generation of mechanical innovations that were already reshaping the workplace. Eventually Wiener himself would have to sound the alarm that his once-obscure theory of cybernation would have untold consequences for workers in factories and offices in a new age of automation.[3]

Robert Wagner and Norbert Wiener traveled divergent paths through the mid-twentieth century, one pursuing reform-minded public policies, the other scientific discoveries. Yet the two are very much connected by this book's major theme. The steady advance of job-replacing technology at the height of the United States' industrial power (1921–66), spurred by the brilliance of innovators like Norbert Wiener, exposed the limits of even the most committed of reform-minded politicians like Senator Wagner. From the New Deal to the New Frontier, an eclectic and fragile liberal-labor coalition was instrumental in enacting new laws that bolstered worker rights, compensated the unemployed, retrained them with new skills, and aided depressed communities. Yet despite its repeated attempts, this ever-evolving amalgam of reformers found itself unable to create a federal economic planning mechanism that might better anticipate the next wave of job-eliminating machines. Successive waves of innovation were consistently unpredictable in how and where they exacted their human toll, leaving politicians to react to damage already done and seemingly powerless to anticipate the technological future. Innovations incubating for years in abandoned warehouses, converted garages, or makeshift laboratories suddenly burst forth for reasons not always predictable—reconfiguring markets, changing cultural norms, ravaging jobs, and hollowing out entire communities. The relentless invasion of new job-replacing technology throughout the period challenged politics to catch up to its effects. While prosperity and an accompanying conservative insurgency also stymied the trajectory of liberal-labor reform efforts, the sheer unpredictability of technological progress ultimately posed the most formidable challenge.[4]

In her meticulously researched work *Inventing Ourselves Out of Jobs?*, Amy Sue Bix explores the complicated debate between jobs and technology from the Great Depression into the 1980s. She describes her own work as "the intellectual history of an idea, the notion that Americans must come to grips with the human consequences of introducing increasingly powerful machines into the workplace." Bix's work continues to occupy a special place between mid-twentieth-century histories of technology, labor, economics, and politics. Understandably, historians of technology such as David Hounshell and Carroll Pursell tell the story of the period's robust industrial development in terms of the increasing sophistication of workplace technology, while labor historians such as Nelson Lichtenstein and Robert Zieger focus on the steady rise of union power and its increasing political agency. When the economist Greg Woirol addressed the history of technological unemployment, he did so by focusing on how those in his field at first theorized that workers displaced by technological unemployment would be reabsorbed by the economy, only to later become more convinced that technological unemployment was a structural feature of modern industrialism. Meanwhile, political historians such as Jefferson Cowie and Julian Zelizer have focused on the achievement of progressive reforms from the New Deal to the Great Society or, alternatively, like Alan Brinkley or Rick Perlstein, on the emergence of the postwar conservative insurgency. Elizabeth Shermer revealed the important role of government-funded technological development as critical to the emergence of sunbelt conservatism. Yet in the broader context, it is Bix's work that continues to fill an understandable historiographical void by tackling the societal impacts of advancing workplace technology and the tensions it created. Extending beyond where Bix left off, this interpretation examines how politics sought to resolve the mid-twentieth-century technology-versus-jobs debate—in particular, how a formidable liberal-labor coalition pushed the envelope of reform yet failed in its ultimate goal of minimizing the jobs-related impacts from successive advances in technology.[5]

The central actors in this narrative are an ever-evolving troupe of political strange bedfellows that emerge in support of New Deal programs, survive in various permutations and combinations through the post–World War II Republican interregnum, and then appear in a reenergized iteration in support of the New Frontier and Great Society. As Meg Jacobs succinctly describes it, "The New Deal had stitched together a political coalition of urban workers, organized labor, northern blacks, white ethnic groups, Catholics, Jews, liberals, intellectuals, progressive Republicans, middle-class families worried about unemployment and old age, and southern whites." Throughout this politically tumultuous period, though this liberal-labor coalition went through its own iterations that strained its very coherence, it remained consistently fixated on technology's impacts on jobs. Such a progressive alignment, though challenged to sustain momentum during the prosperous 1950s, did not relent in pressing the case for worker justice. In addition to Robert Wagner, the evolving coalition included Labor Secretary Frances Perkins, Wagner's longtime friend, who was among the first to warn of the insidious effects technology had on jobs. The New Deal coalition also counted on stalwarts such as Senator Joseph O'Mahoney (D-WY), who chaired a major inquiry into unemployment's causes in 1938, and Senator James Murray (D-MT), who shepherded the Employment Act of 1946. Yet it also housed progressive southern Democrats such as Senator Hugo Black (D-AL), with his thirty-hour-workweek proposal in 1933, and Congressman Wright Patman (D-TX), who a decade after championing full employment legislation chaired the 1955 automation hearings. Nor was such a reform coalition limited to Democrats. It included progressive Republicans like the former Ford executive Senator James Couzens (R-MI), who held extensive hearings on unemployment just a month after Wagner's maiden floor speech. Another New Yorker, the liberal senator Jacob Javits (R-NY), actively sought bipartisan alliances with a new generation of liberal-labor reformers like Senators Paul Douglas (D-MA) and Hubert Humphrey (D-MN).[6]

The coalition's unrelenting concern over those technologically unemployed had its own deeper roots, even though US politics had not only long ignored the loss of jobs to machines, but also walked past the issue of unemployment altogether. As Udo Sautter documents in *Three Cheers for the Unemployed*, prior to World War I, some states, and even the Bureau of the Census, did attempt cursory surveys aimed at counting how many were employed on an industry-by-industry basis. In addition, a smattering of job exchanges arose to help the jobless find work, including a controversial and ineffectual office within the Department of Commerce tinged by nativist politics. Yet, although reformers attempted to generate larger national interest and sympathy for those without work, concern for the unemployed remained limited primarily to religious organizations, local charities, and an assortment of devout reformers. An abiding respect for the dignity of work did not transfer to widespread concern for those out of work. Initiatives that sought to confer on the jobless at least some modicum of political standing, from Coxey's march on Washington to Eugene Debs's plea for unemployment insurance, failed to gain political traction. Moreover, ideas such as Henry George's bold land tax that might allow farm workers increasingly subject to mechanization to share in the value of their labor were overwhelmed by a prevailing social Darwinism. Even American Federation of Labor (AFL) president Samuel Gompers saw unemployment as an affliction of the unskilled. Instead, he turned his attention to improving the hours and wages of those in his growing union of largely skilled workers.[7]

In sum, prior to World War I the federal government scarcely acknowledged the unemployment issue and did nothing to ameliorate its effects. But in the war's aftermath, and prompted by a deepening recession caused by a glut of surplus war goods, Commerce Secretary Herbert Hoover persuaded President Harding that it was time to convene a Presidential Conference on Unemployment. With the United States now the world's leading industrial economy, the government officially recognized that the reality of joblessness could no longer be ignored. Though Harding stacked the gathering with his business allies, the very convening of the conference officially legitimized the government's newfound concern with unemployment and the challenges it posed to policymakers. Moreover, by urging businessmen to become engaged with their communities in finding new work opportunities, and with the establishment of a committee within the Department of Commerce to further study the issue, the federal government took a small, though high-profile, step toward legitimizing the need for government to address this nagging issue.[8]

With the government now on record at the highest level expressing concern for those out of work, it did not take long to connect worries about unemployment to the role that machines were playing in causing it. It was an issue that caught the attention of the newly formed Labor Department and its new secretary, James Davis, who coincidentally had lost his very first job to new technology. Mass production caused an almost overnight transformation of US industry, and throughout the 1920s it fueled a quantum shift in annual productivity gains that would last until the mid-1960s. Henry Ford's willingness to share the techniques he had first adopted at his Highland Park plant, when combined with the science of labor-saving efficiency developed by Frederick Taylor, led to a 23.5 percent increase in manufacturing production between 1923 and 1929. Labor productivity rose some 14 percent over the same period as a new generation of efficient machinery increasingly replaced less efficient workers. The buoyant optimism that greeted technology's advance from the end of World War I through the twenties was suddenly met with the recognition that although technology might be an unqualified blessing for productivity and profits, it was responsible for significant economic dislocations and was a significant contributor to increasing unemployment. Throughout the 1920s a series of scholarly studies, including those done by the future New Deal architect Rex Tugwell, revealed how new factory technologies were displacing workers in selected industries. In 1928, with an article in the *New Republic*, Columbia University professor Sumner Slichter summarized the phenomenon he and many of his colleagues referred to as technological unemployment.[9]

With the introduction of his initial legislative package that same year, Wagner thus commenced the pursuit of a now well-understood phenomenon. He would not have long to wait for the jobs crisis of the Great Depression to inject urgency into the cause. Five years after the enactment of Wagner's bills to simply construct a better count of the unemployed and create a network of job exchanges (1930), the senator's signature National Labor Relations Act (1935) was enacted, guaranteeing labor new rights and, most importantly, allowing for collective bargaining. As a result of this milestone reform, workers might finally begin to realize some share of the enhanced revenues companies derived from the increased productivity that technology made possible. On the heels of Wagner's legislative accomplishment, the Social Security Act was enacted in 1935, requiring employers to pay an unemployment tax, thus bolstering fledgling state unemployment compensation systems. By 1938 the Fair Labor Standards Act finally codified the long-sought-after eight-hour workday, along with allowing for overtime pay and setting a minimum wage. And in the immediate aftermath of World War II, with the experience of full employment fresh in everyone's memory, the liberal-labor coalition sought the enactment of the Full Employment Act (1946), which would place the federal government at the center of insuring everyone "the right to a useful and remunerative job in the industries or shops or farms or mines of the Nation," just as President Roosevelt had promised in his proposed second Bill of Rights.[10]

The postwar debate over the full employment bill marked an important turning point in the politics of the technology-versus-jobs debate. During the Depression, FDR became increasingly convinced that technology was having a corrosive effect on the nation's economic recovery. In the aftermath of the conversion of the economy to wartime footing and the just as rapid and well-executed reconversion to peacetime, the liberal-labor coalition made its most concerted effort to establish a federal economic planning capability that might predict the jobs future rather than allow more workers to be victimized by the next technological turn. During World War II, the impressive mobilization of science and technology, under the direction of Wiener's MIT colleague Vannevar Bush, suggested just how such government planning and coordination might be extended to help ensure full employment in a peacetime economy. The liberal-labor coalition seized on the opportunity presented by the example established by Bush's Office of Scientific Research and Development (OSRD). It was argued that if a command-and-control model directing technological mobilization drove full employment during the war, why couldn't a similar government planning mechanism do the same in peacetime? The OSRD seemed the prototype for the sort of new cooperative public-private partnership that might direct the economy going forward after the war and validate the notion that government could play a prominent and beneficial role in economic planning.[11]

The original Full Employment Act's extensive National Production and Employment Budget was precisely the sort of central-planning mechanism the reformers envisioned as a means to protect jobs by forecasting technology's next turn. But in the war's aftermath, a strained and still-fragile New Deal coalition faced an emergent conservative political movement that strongly opposed any such federal economic planning mechanism. As Donald T. Critchlow describes it, the new postwar conservatism marked a sea change from the crankier prewar Liberty League version, which was simply opposed to the heavy-handedness of the National Recovery Act. The new conservatism had its own intellectual acolytes who were redefining the technology-jobs debate in terms of the vibrant economic effects from government's continued investment in research that they claimed would expand the middle class and bolster the nation's Cold War defense. For the new conservative movement, government planning that created new jobs in those right-to-work Sun Belt states was distinguishable from planning that sought to preserve union jobs in the industrial Northeast. As Lisa McGirr observes in *Suburban Warriors*, "While a broad segment of the southlands business community railed against Washington and resented federal interference, its growing economic power was a product of federal government largesse."[12]

Playing to the suddenly much-enlarged postwar, white, suburban middle class, the leaders of an emergent postwar alliance of southern Democrats and midwestern conservatives effectively neutered the liberal-labor version of the Full Employment Act. They then trained their sights on weakening Wagner's NLRA by enacting their Taft-Hartley amendments. Ira Katznelson has described how after the war, southern Democrats grew increasingly lukewarm to certain aspects of the New Deal, particularly those aimed at worker or racial equity. The government's exploding investment in new defense-related research and development enabled rapid industrial growth in an emergent Sun Belt region with its attendant right-to-work, antiunion bent. As Elizabeth Tandy Shermer notes, the region's dramatic postwar growth, which ironically bore the imprint of New Deal support, presented formidable barriers to expanding any of the earlier liberal-labor reforms. Entering the age of automation in the 1950s, a new generation of liberal-labor reformers found stiff resistance even as they tried to achieve incremental increases in the minimum wage or enact trade-adjustment assistance. Both Senator Paul Douglas's Depressed Areas legislation and Senator Joseph Clark's (D-PA) Manpower Training Act would have to wait for enactment until their former colleague and liberal-labor ally John Kennedy was in the White House.[13]

Yet even as President Kennedy signed these two bills into law, thus seemingly jump-starting the continuance of liberal-labor reformism, his New Frontiersmen grew increasingly concerned over naggingly high unemployment rates. Earlier, on the campaign trail, Kennedy himself seemed to have reached the realization that structural unemployment seemed impervious to the best the liberal-labor coalition had to offer. In his quest for the Democratic nomination, Kennedy spoke often of the toll automation was taking on blue-collar jobs, and he suggested that perhaps a national commission was called for to assist policymakers in getting ahead of the next technological turn. Both Kennedy and his vice president constantly straddled the divide between technology and jobs. Both were strong proponents of technological advancement, as evident in their wholesale advocacy of space exploration. At the same time, at the outset of his senatorial career, Kennedy focused attention on the depressed areas of New England and the fast-dwindling textile industry. Lyndon Johnson had also witnessed firsthand the effects of unemployment on his family and neighbors in his native Texas Hill Country. His War on Poverty would speak to his expanded view of the social costs of joblessness. In seeing through Kennedy's idea by announcing the appointment of a National Commission on Technology, Automation, and Economic Progress, Johnson noted that "technology is creating both new opportunities and new obligations for us—opportunity for greater productivity and progress—obligation to be sure that no workingman, no family must pay an unjust price for progress." Indeed, as the commission met for the first time in January 1965, jobs in the industrial sector had doubled from what they were in 1921. And yet, those on either side of the now four-decade technology-versus-jobs debate were still unable to say with certainty whether technology was creating or eliminating more jobs.[14]

The Technology-versus-Jobs Debate

Throughout the period of US industrial ascendance examined here, this very question became the politically convenient conundrum allowing each side to make claims in the moment and then substantiate them with largely anecdotal information. Politicians conveniently and repeatedly called for task forces, committee hearings, and ultimately a commission that might determine precisely what impact technology was having on employment. As often in politics, perception overwhelmed reality. If, as the techno-optimists claimed, there were more than enough new jobs to be had in emerging new industries, those in the liberal-labor coalition attuned to the plight of workers could credibly argue that jobs, indeed entire skills, were rapidly disappearing and communities suffering as factories shuttered. Senator Kennedy made his first floor speech in March 1953 on the loss of textile jobs in New England, while his friend (and political rival) Senator Barry Goldwater enthusiastically argued for more defense contracts and the jobs created in a suddenly invigorated Sun Belt. Yet what those on either side of the machines-versus-jobs debate shared was an inability to credibly extrapolate beyond their anecdotal examples of jobs lost or jobs created. From an advisory committee appointed by President Hoover in 1930 that attempted to measure how rapidly technological unemployment was occurring, succeeding presidents and Congresses called for one study after another in their quest to answer the question of whether new machines were creating more jobs than they eliminated.[15]

Hoping to finally answer the question, Johnson's Automation Commission produced an entirely new set of analyses that sought to measure the speed at which various industries adopted new technologies, believing it might unlock the policy formula that would head off the next surge of technological unemployment. The liberal-labor coalition hoped that, at last, it would have a basis for planning and thus the ability to mitigate the effects of future technologies on jobs. But although the commission's analysis was more sophisticated than prior efforts, it still did not definitively answer the question. Indeed, a breakthrough in measuring the macro impacts of technologies in creating versus eliminating jobs would not emerge until the 1970s with the Nobel Prize–winning macroeconomic modeling of Jan Tinbergen. Yet even his work was later revealed as less than definitive. Still today, even the cutting-edge analysis of Deron Acemoglu and Pascual Restrepo continues to acknowledge the obstacles to accurately measuring the net impact of the displacement effect of new technologies when they are balanced against their replacement effects (i.e., job creation). In testimony before the House Budget Committee hearings on automation in September 2020, Acemoglu defaulted to the localized rather than national impacts of artificial intelligence on jobs. His testimony seemed eerily reminiscent of similar debates that had raged at the height of US industrial prowess fifty years before.[16]

The inability to ever entirely resolve the displacement-versus-replacement question allowed the most alarmed technophobes to predict a dystopian future with increasing joblessness, while ebullient technophiles envisioned an economy where workers were not only fully employed but freed from drudgery. More often than not in the political debate that evolved over a half century, however, advocates on both sides resisted moving to such extremes. What might seem the predictable dividing lines between business and labor, progressives and conservatives, or Democrats and Republicans were frequently crossed in what emerged as an effort to reconcile the United States' long-held beliefs in both the dignity of work and the value of technology. In speeches and articles in the 1920s, AFL president William Green raised the issue of technological unemployment yet also carefully avoided any critique of technological progress. He set forth a theme that would be consistently invoked by the liberal-labor coalition, that workers should simply be allowed to realize their fair share of the gains accruing from new technology. Walter Reuther, as the dynamic and forward-thinking president of the United Auto Workers, gave the back of his hand to proposals that would ban new technologies from entering the workplace altogether. "Nothing could be more wicked or foolish," he said. Instead he focused on negotiating agreements such as the breakthrough Treaty of Detroit in 1949 that, consistent with Green's earlier admonition, allowed workers their fair share of productivity gains made possible by new technology.[17]

Nor, by the same token, was the automation guru John Diebold, retained by scores of Fortune 500 firms, callous toward the immediate human consequences of computer-driven factory lines. In his 1952 book titled simply

Automation, Diebold persistently made the case that although the use of computers could cause workers in specific industries to lose jobs, its net effect
was to raise the quality of work and liberate the worker. Humans, Diebold
maintained, were inefficient in carrying out industrial tasks. Ultimately he
argued that the new automatic equipment would create a burgeoning new
service economy, with workers increasingly less subordinate to the pace of
workplace machines. With an almost naive optimism, he asserted that "the
worker will be released for work permitting development of his inherent
human capacities." In his testimony at Congressman Patman's extensive 1955
automation hearings, Diebold suggested how an automated future would
give rise to more competitive enterprise, a robust new service sector, and a
general upgrading of jobs. And yet, at the same time, just as labor leaders
treaded carefully in denigrating technological progress, Diebold laid out an
extensive blueprint for a comprehensive analysis that might anticipate automation's social as well as economic impacts, and he readily acknowledged
that government needed to play a larger role in accelerating the retraining of
those workers displaced by technology. Indeed, the fiercest of advocates on
both sides of the technology-and-jobs debate recognized that both employment and technological progress were valued components of the American
ethos, and thus assiduously avoided any wholesale dismissal of one or the
other.[18]

Indeed, throughout its four-decade political crusade, the liberal-labor
coalition consistently sought to tame the excesses of workplace technologies, not eliminate them. Carl Benedikt Frey documents how the contentious
technology-versus-jobs debate has often been characterized by conflicted
worker attitudes toward technology. In those periods where workers feel
economically threatened by new machines, or what Frey terms a "technology trap," they tend to overlook improved living standards and mount
attempts to slow the pace of technological advance. Here, however, it
will be argued that in their representation of the worker during the mid-
twentieth century, the liberal-labor coalition, through both good economies
and bad, was more devoted to anticipating technology's next turn than it
was to discouraging its advance. To be sure, Reuther hoped his idea of an
expanded guaranteed annual wage, which forced auto companies to devote
some monies to a fund for laid-off workers, might also serve to slow the
pace of capital expenditure. But unions were generally lukewarm to the
idea of any machine tax. Put in perspective, the liberal-labor coalition's crusade for the unemployed focused on ensuring that workers received a fair
share from technology's contribution to improved productivity, while at the

same time seeing to it that the unemployed were well compensated, fully retrained, and shepherded toward new work.[19]

Seen through the prism of the technology-versus-jobs debate, the interpretation that unfolds in the chapters that follow largely comports with Jefferson Cowie's argument that the achievements of the liberal-labor coalition constitutes "the great exception." In their original iteration, which presented a perspective on the impressive trajectory of mid-twentieth-century progressive successes, Cowie and Nicholas Salvatore distinguish the Great Society's view of the poor as the victims of systemic societal problems from the earlier New Deal perspective that saw government's role in job creation as a remedy for those harmed by the collapse of capitalism. Cowie notes how the targets of the War on Poverty were those falling outside the coverage of a labor union movement and the already strong set of benefits it successfully negotiated for its members. Cowie in fact contends that although "Johnson was of the New Deal, the Great Society was not." Indeed, Johnson did summarily dismiss the need for a separate jobs program, opting for a Jobs Corps that focused its retraining on the young and minorities. Moreover, his War on Poverty is known more for programs aimed at the poor, such as Head Start and Medicaid, than for improving the condition of the unemployed. As the major cog in the liberal-labor coalition, unions by the 1960s were exacting greater benefits at the bargaining table than they were in the halls of Congress. Now operating autonomously to preserve an ascendant blue-collar middle class, unions could no longer be counted on to be at the center of the liberal-labor coalition, which itself was listening to the fresh voices from a New Left movement. In her description of the boundaries placed around jobs policy post–New Deal, Margaret Wier describes how the stiff headwinds that the liberal-labor coalition encountered in the sixties, exacerbated by racial and populist fault lines, soon moved many rank-and-file union members into Nixon's silent majority. Indeed, by the time Johnson's Automation Commission convened in January 1965, the union movement was undergoing a transformation that would later manifest in a backlash to the new left turn in liberal-labor reform, a turn dramatically revealed in the Hard Hat Riot on Wall Street in May 1970.[20]

Yet it is possible to see how, even at the outset of this transitional moment as the Automation Commission met, the liberal-labor coalition remained committed to seeing through reforms that addressed the impacts of technology on blue-collar workers and their communities. The impressions left on Kennedy as he met anxious miners coming out of the coal mines during his frenzied West Virginia presidential primary campaign caused him to begin actively talking about the insidious effect of automation on jobs. And in the early days of his presidency, he signed the bills that Senators Paul Douglas and Joseph Clark had each doggedly advocated in the fifties, despite fierce conservative opposition and Eisenhower's tactical resistance. These liberal-labor reformers, with whom Kennedy had worked side by side during his Senate career, persisted in carrying the unflagging liberal-labor commitment to help retrain the unemployed and rehabilitate their depressed communities. As he had on the campaign trail, Kennedy also asked Congress to create a commission to design more creative solutions to the challenge of structural unemployment. In carrying Kennedy's idea forward, Johnson, spurred on by his friend Reuther, insisted that a commission could possibly deliver consensus recommendations to create a federal planning agency that might anticipate the technological future.

In delivering its final report to Congress in February 1966, the commission noted, "As a nation we have willingly accepted technological change, but we have never been fully successful in dealing with its problems even when the pace of technological advance and the growth of the labor force were less rapid than today." The report might well have added that this deficiency was precisely why the commission had been formed in the first place. Acknowledging that the commission was no more clairvoyant than previous efforts, one of its members, Daniel Bell, who in 1973 would ordain "the coming of post-industrial society," fondly recalled the French philosopher Bertrand de Jouvenel's insight, "How can the future become a matter of public opinion?" Yet Bell may have understated the commission's many breakthrough contributions. In designing its own innovative approaches to the automated future, including its call for a basic income and public education through grade fourteen, the commission seemed to move beyond the liberal-labor coalition's fixation with economic planning. Finally, a bold set of policy recommendations seemed as innovative as the technologies that necessitated them. Yet they were issued at a politically inopportune moment, and the liberal-labor coalition, so long stymied in its attempt to construct a means to anticipate the technological future, suddenly seemed just as incapable of embracing bold measures that spoke to it.[21]

The issuance of the commission's report marked a fitting endpoint to the midcentury technology-versus-jobs debate. Economically, four decades of consistent productivity gains, which persisted even through the Great Depression, were coming to an abrupt end. The golden age of US manufacturing that witnessed nearly a doubling of jobs in the industrial sector from 1920 to 1970 was also beginning to plateau as it faced stiff new foreign competition. By the early 1970s, inflation and energy supplies became the new bogeymen costing jobs in a new recessionary climate, and automation was eclipsed as the culprit causing unemployment. Politically, by the end of the decade, the nation's attention was fixated on an escalating war in Southeast Asia and accompanying social and racial tensions. The distant echoes from the liberal-labor coalition's desire to mandate full employment and create a centralized federal planning function would emerge in one last hurrah for the supporters of full employment legislation in the late 1970s. Yet during the debate over what resulted in the tepid Humphrey-Hawkins amendment to the 1946 act, technology, much less automation, was hardly mentioned at all. A strained liberal-labor coalition aimed its sights instead on ending the war, enabling civil rights, and restoring the environment. The commission's textured and far-reaching response to technological advancement seemed irrelevant to the politics of the moment.[22]

However, as the information economy of the twenty-first century matures, the jobs-versus-technology debate has again begun to take familiar turns. Our current politics is challenged to ensure that anticipated new advances in artificial intelligence and quantum computing will not have devastating results for industries and regions, and that the latest contortions of our information age will not reshape the contours of our economy or even our politics with disastrous social or political consequences. Moreover, the past suggests it is the technological breakthroughs like Norbert Wiener's cybernetics, conducted away from the glare of public view, that will prove the most challenging. In such an unpredictable context we are reminded of the inherent randomness of technological discovery. As David F. Noble points out, there is no "technological Darwinism," no theory that predictably sorts between competing discoveries or renders politics the innocent bystander in some grand technological determinism. As much as the government proactively funded the development of new technologies, particularly after World War II, it was not capable of predetermining the eventual technological outcome, much less the social and political fallout from its own investments. Few better examples exist of technology's random course than the government's relatively small $125 million investment to connect its own computers in what became known as ARPANET, the precursor to the World Wide Web and the basis for unleashing a multitrillion-dollar information economy.[23]

As we contemplate where technologies of the twenty-first century may lead, many of the solutions being offered sound similar to those offered by the liberal-labor coalition as it chased after the unforeseen consequences of technologies in the mid-twentieth-century industrial United States.

Understanding the limitations that previous reformers encountered should temper our optimism about what is politically possible. Yet their story may also serve as an inspiration. As the following pages reveal, the significant accomplishments of these determined reformers were achieved because of their steadfast commitment to goals that seemed unreachable when they began, at a time when the unemployed were not counted and politically did not count.

CHAPTER 1

Voices for the Unemployed

The rapidity of our inventions and discoveries has intensified many problems in adjusting what we nowadays call technological unemployment.

—Herbert Hoover, speech to the American Federation of Labor, October 6, 1930

Growing up in the heart of iron and steel country in Sharon, Pennsylvania, James Davis decided to follow in his father's footsteps. He became an apprentice iron puddler and in 1891 left his hometown to find his own furnace in the mills of nearby Pittsburgh. There he soon began shaping hot molten iron as it streamed from the furnace. Iron puddling was backbreaking work, but it paid well. The iron making that employed Davis relied on a combination of patents obtained by the eighteenth-century ironmaster Henry Cort. The brilliant British innovator showed how, with certain chemical additions and process modifications, iron making could produce a less brittle "wrought iron" that could be rolled into railroad ties, bridge beams, and supports for ever-taller buildings. Yet as happened with so many technologies during the United States' industrial transformation, wrought iron was quickly eclipsed by the new technology of steelmaking. Improvements in the Bessemer process made steel the metal of choice, and suddenly iron puddlers were not in much demand. Facing an uncertain future, Davis decided to move to Indiana, where he became a city clerk and began a new career in politics. Advancing technology made Davis's job obsolete, as it would millions of others in the mid-twentieth century. In his new job, Davis would soon find himself having to account for what would be called "technological unemployment," a phenomenon he had already experienced firsthand.[1]

On August 16, 1921, the "iron puddler" James Davis, by then the secretary of labor under President Warren Harding, submitted to Congress his department's report on what appeared a fast-growing unemployment crisis. The nation was suffering a deep postwar recession; some already were calling it a depression. The coordinated mobilization of the economy entering World War I, skillfully led by the New York financier Bernard Baruch, had given way to a less efficient postwar demobilization. The glut of war products that had moved onto the market drove prices downward. A haphazard transition was marked by agricultural overproduction, significant inflation, and a proliferation of labor strikes. Unemployment seemed ubiquitous, yet no one in authority could say exactly how many were out of work.[2]

Davis's report responded to Senate Resolution 126, which had been offered earlier in the month by Labor Expenditures Committee chairman Senator Joseph McCormick (R-IL). The grandson of the *Chicago Tribune* owner Joseph Medill, McCormick had tried stints with his grandfather's newspapers, including as reporter, publisher, and even owner. Yet he was attracted by politics and won seats as a progressive Republican in the Illinois House in 1912, the US House in 1916, and the Senate in 1918. He dedicated himself to progressive causes, including the unemployed and the government's basic inability to render an accurate count of those without work. His brief resolution simply called on the Department of Labor "to advise the Senate as to the estimated unemployment in the several states, including the number of men, of ex-service men, and of women, estimated to be unemployed." The resolution quickly passed, and Secretary Davis was asked to provide Congress with at least a preliminary answer. In assembling its numbers, Ethelbert Stewart, the commissioner of labor statistics, relied on selective surveys that measured how many workers occupied jobs in various industries and in agriculture as compared to the previous year. Estimates of unemployment ranged widely from two million to six million. Davis erred toward the high side, candidly reporting to the Senate that 5,735,000 more workers were unemployed than when Harding took office. The Labor Department, however, was hardly equipped to make an accurate unemployment census, and the statistics Davis provided were at best rough estimates. The US government remained unable to accurately estimate how many were without jobs.[3]

Meanwhile, New York City congressman Meyers London, the sole Socialist Party member in the Sixty-Seventh Congress, was scolding his colleagues for their neglect of the unemployed. Though at the time Meyer's alarm seemed politically hyperbolic, Secretary of Commerce Herbert Hoover sensed a larger public frustration over the extent of joblessness in the postwar economy. He began lobbying President Harding to convene a conference to explore what could be done. The president agreed, and then quickly asked Hoover to chair what was officially called the Presidential Conference on Unemployment. As the inveterate engineer, Hoover considered no problem to be beyond rational analysis applied by thoughtful experts. He was fresh from his successful leadership of a voluntary cooperative initiative that saved a starving Belgium during the war, and he believed that the same sort of voluntarist spirit might help pull the nation out of its deep recession. In his introductory remarks at the quickly convened conference, Hoover declared, "We possess the intelligence to find solutions. Without it our whole system is open to serious charge of failure." Hoover's social philosophy was based on his optimism that individuals working cooperatively could achieve social good with government support rather than interference. He suggested to the many businessmen assembled that they consider voluntarily shortening workweeks, even though doing so might require workers to accept reduced pay. He also urged the conference attendees to return to their communities and engage their own local governments in hometown partnerships to put the unemployed back to work. Whatever solutions they devised for their own communities, Hoover exuded confidence that individual business own-ers and their local officials could work together cooperatively to alleviate unemployment without Washington's help.[4]

Harding struck an altogether harsher tone. His remarks reflected the still-prevalent social Darwinist view of the unemployed as a class of individuals who were responsible for their own miserable condition. Though the nation lacked any effective social safety net for those out of work, Harding was adamant that there be none. The president instructed the conferees to make certain that their recommendations did not include any federal support for what he considered a core group of individuals who failed the Darwinian test. As he told the business-dominated conference, "There is always unem-ployment. Under most fortunate conditions I am told there are a million and a half in the United States who are not at work. The figures are astounding only because we are a hundred million, and this parasite population is always with us."[5]

The president had little desire to see any money or resources devoted to those he considered unable to fend for themselves. If James Davis could rise to become his secretary of labor after losing his job as an iron puddler, then certainly others should be able to find gainful employment. Populated by only a few labor representatives, the conference recommended duti-fully delegating the unemployment issue back to the states and localities for resolution. Consistent with Hoover's directive, businesses were asked to identify employment opportunities in their own communities, strengthen local charities, and make sure families provided for their own well-being. In essence, and consistent with the still-prevalent laissez-faire notions of business responsibility, the unemployed were left to fend for themselves. Despite the resistance of its namesake, however, the Harding Conference did provide demonstrable public recognition that in an industrial economy the unemployed deserved attention and needed assistance. Though the deep but brief postwar recession would end, and the US economy would prosper as never before, labor leaders, academics, and a nucleus of reform-minded politicians also began focusing on the role technology was playing in displacing workers from their jobs. By decade's end, these early seeds planted out of concern for the unemployed would take root during an unprecedented unemployment crisis.[6]

The Roaring Economy

Fortunately for the Harding administration, even as the conference met, there were already ample signs that the economy was beginning to recover from the brief Harding recession. Jobs seemed to be more plentiful, and by the beginning of 1922, the economy began to roar like the factory floors themselves. The decade of the twenties would be defined economically by prosperity and politically by Republicanism. The plight of the unemployed, hardly recognized before the Harding convention, again largely disappeared from US politics. As the newly elected President Coolidge remarked, "The man who builds a factory builds a temple; the man who works there worships there." While a few remaining progressive Republicans from the Roosevelt-Taft era, like Senator Robert La Follette (R-WI), continued to display their reformist zeal, the reform spirit within the Grand Old Party faded quickly after the war. Successive Republican Congresses during the 1920s were defined by the conservative leadership of House Speaker Nicholas Longworth (R-OH). At the beginning of the 69th Congress (1925–26), Longworth became so annoyed with the twelve Republicans who supported La Follette in the 1924 presidential election that he denied them entrance to the Republican conference and stripped them of their committee assignments.[7]

This conservative political discipline seemed to mirror the US industrial economy, now the largest in the world, that was defined by Henry Ford's assembly line efficiencies, rapid mechanization, higher wages, reduced prices, and mass-produced goods. In the twenties, technology seemed joined to economic progress and not political progressivism. From the tractors on the farm to the new Model Ts on freshly paved highways and city streets, all made possible by new factory equipment that often substituted for manual labor, new technologies permeated American life and spurred robust economic growth. Traversing the ocean in his one-engine airplane in 1927, Charles Lindbergh became a living symbol of the United States' technological prowess and individual resourcefulness. Thomas Edison's research complex in Menlo Park, New Jersey, seemed emblematic of the nation's inventive genius and technological savvy. Popular literature was filled with the can-do spirit that seemed to mark America's inventiveness. As a story in *Collier's* magazine proclaimed, "You don't have to be an inventor to know we're on the threshold of an age of wonders made ordinary, which will bring health, wealth, and happiness to each of us."[8]

The accelerating growth of the industrial economy was powered by a dizzying array of new labor-saving machinery in the workplace. In the auto industry, skilled workers operated machines that assembled car frames with virtually no human intervention other than to operate the machines themselves. Two hundred workers who might be able to assemble thirty-five frames a day under older piecework processes could now turn out some nine thousand in the same period. Between 1926 and 1930, the number of employees needed to operate the Bessemer steel process decreased by 24 percent. In total, the American Federation of Labor (AFL) estimated that in the decade 1919–29, the nation's factories produced 42 percent more goods with 585,000 fewer workers. One company could produce an astonishing 2,000 radio tubes every hour with only three operators, in comparison to the forty workers it previously took to produce a mere 150 tubes. Even in older industries, machines were fast eliminating jobs. In the boot industry, one hundred machines took the place of some twenty-five thousand workers. In 1913, one person could turn out five hundred razor blades in a day. Now with new machinery, that same person could turn out thirty-two thousand blades.[9]

Driving such remarkable growth was the widespread adoption of Henry Ford's approach to industrial organization. During the twenties, the efficiencies of Fordism permeated the ranks of US industry and came to be equated not only with its emphasis on assembly lines and mass production, but also with accompanying higher wages and lower prices. Ford insisted his workers be able to afford the cars they produced, and his methods were increasingly adopted and widely praised. In 1925, Edward Filene, a successful and progressive-minded Boston retailer, gave Ford much of the credit for the

United States' economic ascendance. At its core, Filene observed, Fordism not only meant more efficient production, but also higher wages and reduced prices. As a retailer, Filene knew the positive impact that Fordism had on his own thriving department stores. "Mass production is therefore production for the masses," Filene observed. Fordism was causing other businessmen to recognize the need for wage growth, reasonable hours, and reduced prices, at least in theory if not always in practice.[10]

As the decade wore on, machine-induced improvements to traditional processes—or, in many cases, the introduction of new industrial processes altogether—took a toll on skilled workers of every type, from iron puddlers to farmhands. Labor spokesmen suggested that manufacturing production increased 45 percent between 1919 and 1929 and that productivity rose 53 percent in that same period. Subsequent analysis has shown their observations to have been close to the mark. The advocates of technological advancement during this period readily acknowledged that such dramatic increases in productivity were costing jobs. Following the brief Harding recession, however, these concerns remained a subject for speculation largely among academics, public intellectuals, and labor leaders. The increasing numbers and varieties of products sold at reduced prices, the general increase in wages, the growth of installment buying, and the unfettered stock market speculation combined to fuel an economic expansion that lasted through the decade. With so many new jobs being created, it was easy to ignore those being lost. It remained a steep political challenge to make the case that the technology helping fuel the Roaring Twenties was also creating a new class of technologically unemployed. Yet a few quiet voices began to take up this cause.[11]

A Steady Voice for Labor

Throughout the 1920s, labor leaders focused primarily on a shortened workweek, improved working conditions, and payment of just wages for their dues-paying members. But by 1927, the AFL's new president, William Green, began making the case that businesses should pay a social wage that reflected the amount of profit resulting from new labor-saving machines. Green had entered the coal mines of Ohio as a teenager and became a lifelong union member, rising steadily in the ranks. He ascended to the head of the AFL in 1924 having chaired its committee on a shortened workweek. Conciliatory and cooperative, he was less a public figure than his predecessor Samuel Gompers, or for that matter many other labor leaders whose colorful personalities punctuated the union movement during Green's twenty-eight-year tenure. Nonetheless, despite his almost bookish manner, Green was a tough and skilled negotiator. He adroitly advanced the union agenda throughout the later New Deal era and saw through the postwar merger of the AFL with the Congress of Industrial Organizations (CIO). Moreover, his early advocacy for allowing workers to share the benefits of enhanced productivity would become a theme repeated in different forms by labor leaders in the years ahead. Walter Reuther would successfully expand on Green's idea by negotiating a guaranteed wage that might help deter future layoffs.[12]

Green's evolving thinking on the impacts of machines, first focusing on how workers should gain from the enhanced productivity made possible by automatic equipment and then later suggesting how machines were increasingly displacing workers, was a direct challenge to the Coolidge administration's more sanguine outlook on new labor-saving machinery. Ethelbert Stewart, the commissioner of the Bureau of Labor Statistics (BLS), argued that since laws tightening immigration reduced the domestic labor supply, industry needed to take full advantage of new technology. He contended that industries in the North too often avoided making themselves more efficient by relying on cheap immigrant labor, just as southern businesses clung to what he described as "cheap negro labor." In an explicitly nativist essay entitled "Wastage of Men," written three months after the passage of the highly restrictive Immigration Act of 1924, Stewart reflected the administration's anti-immigrant bias. Dependence on cheap immigrant labor was, in his view, perpetuating inefficient practices, placing more men in jobs that lacked meaning, and ignoring the potential of labor-saving machines.[13]

While Green was not averse to immigration restrictions to protect his own membership, he also argued that accelerating capital investment in labor-saving devices undermined union efforts to negotiate for shorter working hours and higher wages. Throughout the twenties, despite industry's general lip service to a shorter workweek, the adoption of an eight-hour day remained sporadic. Green did argue, however, that one of the benefits of shortening the workday and standardizing a five-day workweek would be reducing the need to displace more workers with new machines. In effect, work would be shared across a larger labor pool if the average hours per worker were reduced. At the same time, Green continued to advance the argument that workers should share with management in the increased profitability afforded by more efficient production resulting from the introduction of new technology. In short, while seeking shorter work hours and better pay, Green was willing to accept the inevitability of advancing technology. In fact, in the context of the Roaring Twenties, he agreed that machines coupled with shorter work hours did not necessarily have to mean fewer jobs. Moreover, what the new machines should allow, Green argued, was improved pay for his membership.[14]

Yet as the decade wore on, Green became increasingly reconciled to the reality that advancing technology was indeed creating a class of what he began referring to as the "technologically unemployed." The labor economist Sumner Slichter, in an article in the *New Republic* in 1928, gave legitimacy to the idea, writing, "Clearly we are confronted with what appears a new kind of unemployment problem—the problem of unemployment caused by technical progress emerges clearly into view." Slichter was among a group of economists who began to challenge conventional, though still unproven, assertions that overall net employment growth accompanied technological change. In his public utterances outside the cramped and smoke-filled negotiating rooms where labor and management exchanged stinging barbs, Green began to echo Slichter's alarm at the increasing rate at which new machines displaced workers in factories and on farms. In a Labor Day article in 1929, just weeks before the stock market crash, Green claimed that the combination of automatic processes, new machinery, and industrial methods was responsible for increasing productivity 53 percent since 1919. Yet he argued that such spectacular gains in efficiency came at the expense of tens of thousands of workers displaced by machines or denied employment. A later article he wrote for the *New York Times* in 1930 warned that the creation of jobs in the new service industries (e.g., gasoline stations, auto repair shops, hotels, restaurants), and the growth of jobs in the professions (e.g., teachers, doctors, lawyers), could not offset the sheer numbers of new workers entering the workforce, and those jobs being eliminated by machines.[15]

The Senator for the Unemployed

As Green began advancing the idea that the new workplace technology was expanding the ranks of the unemployed, the Republican-dominated politics of prosperity continued to focus on tightening immigration, hiking tariffs, and prohibiting alcohol. Progressive reform measures enacted during the Roosevelt-Taft era remained underfunded, and government regulation of industry, as thin as it was, further scaled back. Meanwhile, the unemployed remained beholden to a patchwork of private charities, church relief funds, and local government leftovers. Not surprisingly, then, as the New York City native Robert Wagner arrived in the Senate in 1927, he told his colleagues that those who dared speak about the unemployed were looked on as alarmist. But the new senator wasted no time proposing legislation that would first give the unemployed standing and then give them relief. Wagner's progressive roots dated to his leadership of the New York State Factory Investigating Committee, formed to probe the causes of the devastating Triangle Shirtwaist Factory Fire of 1911, the greatest industrial tragedy in New York history. The commission's recommendations led to a series of measures that set new workplace health and safety standards. Working with his mentor Al Smith and his progressive colleague Frances Perkins, Wagner gained a reputation as a reformer concerned about the United States' working class, which catapulted him to majority leader of the New York Senate and on to the US Senate in 1926.[16]

Consistent with a timeworn tradition, the freshman senator waited until shortly past noon on March 5, 1928, a year after he took his oath, before he gave his first floor speech. It was seven years since Senator McCormick's failed effort to establish a thorough counting of the unemployed, and still there remained no accurate assessment of unemployment numbers. Despite what seemed a still-booming economy, the BLS reported a 15 percent decline in employment based on its own routine surveys of some selected manufacturing facilities in specific industries. Skeptical of the administration's inexact numbers, Wagner believed the unemployment situation was much worse. He offered his own simple resolution, not dissimilar to the one McCormick had offered seven years before, or, for that matter, the one that his fellow senator David Walsh (D-MA) had offered the year before. In fact, the measure simply underscored a conclusion of the earlier Harding conference report that "the first step in meeting the emergency of unemployment intelligently is to know its extent and character." Wagner's resolution required the administration to investigate and compute the number of unemployed, rather than simply extrapolate from surveys of specific industry employment rates. Importantly, recognizing that current methods of counting the jobless were inaccurate, it stressed that the administration should report on exactly how future computations might be improved. The unemployed represented millions who had no other means of expression, Wagner concluded. The least that could be done was to begin with an accurate count.[17]

During a booming economy, Wagner's colleagues did not seem to share the freshman senator's alarm. Surveys of the economy for the previous month indicated that despite some sluggishness, "Coolidge prosperity" maintained its dizzying momentum. Though there was some anecdotal evidence of a softening jobs market, the nation's banks were reporting strong conditions and most industries were doing well. Wagner's anxieties and warnings were judged out of place in the still-thriving economy. The *New York Times* reported that the state's junior senator simply was making "a major political move in the interest of Governor Smith's boom for the Presidency." Despite the political criticism, Wagner's Senate Resolution 147 asking for an investigation and computation of the unemployed was innocuous enough that, in what appeared a painless gesture of senatorial courtesy to the freshman member, it passed by unanimous consent.[18]

In response to the resolution, Coolidge's BLS took but a few weeks to assemble its findings. On April 30, BLS commissioner Stewart confidently reported to Congress that as of March 26, 1928, the precise number of those out of work was 1,874,050, not the some four million that Wagner had claimed in his maiden Senate speech. Nonetheless, the BLS number did not seem to square with what people were beginning to observe all around them. And after a month of carefully analyzing Stewart's numbers, Wagner took to the Senate floor again. His blistering critique spared no punches. Wagner claimed Stewart's numbers were not simply misleading, but inconsistent with the administration's own reports on employment. For example, the Agriculture Department had earlier reported three million people leaving farms between 1921 and 1927. Wagner rhetorically asked whether all these farmworkers were absorbed into industrial jobs. Meanwhile, the Labor Department reported two million new young people entering the workforce over the same time frame. Wagner cited the department's claim and then compared it with the number of new jobs it claimed were available nationally. The two did not square. Wagner's relentless critique concluded that if he was correct in his estimates, at least four million workers were out of work. If his methodology was flawed, it was no more so than the haphazard BLS approach. Either way, the country simply lacked a system for more accurately estimating the numbers of unemployed.[19]

Wagner's Three Bills

Disappointed that the Coolidge administration did not take his earlier resolution seriously, Wagner decided to introduce three separate bills to address what he perceived as a growing unemployment crisis. In his lengthy introductory remarks, Wagner cited estimates that new technologies alone were principally responsible for a 29 percent increase in productivity in the period from 1919 to 1926. From his staff's meticulous analysis of selected industries, and with assistance from the American Statistical Association (ASA),

Wagner noted that when extrapolated across all industries, such a gain in productivity would mean a 7 percent decline in the number of jobs available. The trend could not continue without significant consequences, and the unemployment problem clearly needed to be better understood. Wagner asked the experts at the ASA for advice on how the BLS survey could be expanded to other sectors of the economy, allowing for a more representative sample. Thus, the first of Wagner's three bills proposed that the BLS receive $100,000 to consult with the ASA and other national experts in order to improve its own unemployment analysis. S. 4158 mandated that, based on these consultations, the BLS then estimate and report each month the number of unemployed. In doing so, under the requirements of Wagner's bill, the bureau could no longer be allowed to base the estimates simply on anecdotal surveys of those with jobs in a few select industries. It would now have to make a comprehensive survey of all industries and all sectors of the US economy in order to obtain a much clearer assessment of unemployment nationwide.[20]

Wagner's other two bills addressed the needs of what he described as the technologically and cyclically unemployed. The first (S. 4157) called for the establishment of a network of state employment exchanges under the guidance of the Department of Labor's Employment Service. Roughly half the states possessed some form of public employment bureau, yet their performance was inconsistent. Wagner envisioned an interconnected national network of separate statewide offices that would focus on the needs of that growing segment of the jobless whose jobs were being replaced by machines. Wagner described how new machines introduced into the workplace forced workers to change occupations and locations, invariably navigating the process themselves. Meanwhile, he noted, other industrialized countries already possessed such job clearinghouses. Therefore, Wagner called for the establishment of a federal employment center dedicated to matching the unemployed with available jobs in other states by networking information between state offices. In an economy where skilled jobs could vanish overnight with the introduction of the next new machine, a system was needed for workers to be retrained and placed in jobs where work was available.[21]

Though skills training was a constant default response to technological unemployment, the third bill in Wagner's triumvirate, the Unemployment Emergency Act of 1928 (S. 4307), established a foundation for what became the aspirational goal of the liberal-labor coalition. Over the next four decades, various attempts would be made to establish a federal forecasting and planning capability that could anticipate technological advances and ameliorate their impacts on jobs. In this first iteration, Wagner proposed the creation of a Federal Employment Stabilization Board (FESB). As envisioned, the independent board would comprise the secretaries of labor, commerce, agriculture, and the treasury. The board was to advise the president on trends in unemployment, allowing for timely forecasts of forthcoming economic downturns. Importantly, with the consent of the president, the board also possessed the authority to escrow a portion of federal budget surpluses in prosperous years that in turn could provide for expenditures on public works during economic downturns. This new planning mechanism was to take advantage of what Wagner hoped would be the improved unemployment census called for in S. 4158. If the board foresaw within a six-month horizon any area of business depression and unemployment, then it could advise the president in order that a request for supplemental appropriations be submitted to Congress requesting public works projects. Wagner made the case that these projects not only would provide jobs, but also would stimulate the purchase of supplies from business to spur economic growth.[22]

As Wagner concluded his lengthy floor remarks introducing his legislative package, he asked his colleagues bluntly, "What are you going to do about unemployment?" In 1928 most members could politically afford to sidestep that question. The politics of prosperity had no time for concerns about the jobless. Moreover, since the mid 1890s, except for Woodrow Wilson's presidency and an occasional majority in one or the other congressional chamber, the Democrats were the minority party. In addition, southern Democrats, in a portent of what was to come, were typically hesitant to endorse northern Democratic reform efforts. In the 70th Congress (March 1927–March 1929), the Republicans maintained their firm majority grip on both chambers as they had done throughout the decade. Wagner's three bills were not going anywhere. The best the Republican Congress could muster in addressing unemployment was to pass an amendment offered by their progressive colleague Robert La Follette Jr. (who had replaced his father), appropriating more money for the BLS to continue its data collection methods that inadequately measured the extent of unemployment.[23]

Eight months before the market crash, however, the Senate Labor Committee did issue a report on the problem of the growing unemployment phenomenon already becoming evident. As the decade ended, there seemed a growing anxiety over rising joblessness and the role that new technology was playing. "Machinery and discovery are every day displacing men whose lives have been spent in developing the skill and ability necessary to their crafts," the committee report concluded. The report also acknowledged that "technological unemployment" was spreading rapidly. Following Slichter's lead, a new consensus was emerging among economists that due to the steady incursion of advanced technology in the workplace, cyclical unemployment was no longer adequate to explain what appeared to be a growing unemployment crisis. Though some new industries would always be able to absorb some of those thrown out of work, a larger human toll was being taken by the incursion of new technology in the workplace. Thus, entering the 71st Congress in March 1929, Wagner reintroduced his three bills and stepped up his campaign to enact them. He promoted his legislation with radio addresses to make his points to a wider audience. Speaking to those who might be out of work, he said that he wanted to give voice to "the unemployed who do not give campaign contributions . . . nor maintain any lobby in Washington to tell their depressing story." His three bills seemed to have met their moment.[24]

A Gathering Consensus

Though Wagner remained the most persistent congressional voice on behalf of the unemployed, he was not alone in sounding this early alarm. Senator James Couzens (R-MI), chairman of the Senate Labor Committee whose report warned of the growing problem, presided over twelve days of hearings on the unemployment situation in December 1928, devoting particular attention to the idea of establishing a system of unemployment compensation. Prior to becoming a senator, Couzens was among the original investors in the Ford Motor Company and one of its first officers. When he left the company, he converted his initial $7,000 investment into a $30 million return. Couzens defied the stereotypical 1920s Republican laissez-faire view of business. As one observer remarked, Couzens "was never a regular millionaire." An enlightened philanthropist, he made a gift of $10 million to establish the Children's Fund of Michigan. In his new political life after Ford, he first served a term as Detroit's police commissioner and then became its progressive Republican mayor, attempting a sort of civic welfarism. Betraying his own early experiences working on the railroads and later in the coal industry, Couzens was attuned to the effects of unemployment and attendant poverty as having large societal costs. Joined by other progressive urban business leaders, he sought dramatic increases in the city budget for public works improvements to raise living conditions. His reputation for civic-mindedness as mayor carried over to the Senate as he began raising concerns over technological unemployment. Following the Labor Committee's extended hearings on unemployment, its report acknowledged that many workers, through no fault of their own, were being displaced by machines. As the hearings concluded, Couzens posed the question, "What becomes of these men? What can be done about these thousands of individual tragedies? What do these individual tragedies mean to society as a whole?" As the decade ended, there were still no concerted policy responses, and few politicians like Couzens or Wagner who challenged their colleagues to address a growing problem.[25]

Often those speaking up on behalf of the growing constituency of the unemployed remained on the political margins. The fifth-term Milwaukee congressman Victor Berger was a proud member of the Socialist Party. After he won his congressional race in 1918, Congress denied him his seat, accusing him of violating the Espionage Act. Yet, like Wagner, he was already moving beyond simply recognizing the unemployment problem and proposing to do something about it. His unemployment compensation bill introduced in 1928 was modeled on European plans. It required employers, employees, and the federal government to pay into a national fund. Berger accepted that unemployment was the inevitable and inescapable condition of the industrial system. Being without a job was a reflection neither on the individual nor on society. He estimated that even in a period of relative prosperity, as many as ten to twelve million had inadequate incomes. Berger knew he was ahead of the politics of the late 1920s; six states had already rejected state unemployment compensation fund proposals. He nonetheless expressed confidence that sooner or later his plan would be adopted.[26]

Couzens courageously followed suit, offering his own proposal for unemployment compensation. More modest than Berger's, Couzens's plan sought to build on the reality that some companies were already stepping up their efforts to attend to the immediate needs of the employees they laid off, even as they introduced new labor-saving equipment. Yet these volunteer corporate welfare programs exhibited little consistency in how much they set aside or how they dispersed benefits. The well-known progressive-minded University of Wisconsin economics professor John Commons advocated a nationwide unemployment compensation system that paralleled his own state's workmen's compensation system, which required employers to pay into a state-backed insurance fund. Under the Commons plan, companies would be required, rather than simply allowed at their discretion, to set aside a specific amount. Couzens's Labor Committee conveniently seized on Commons's testimony to help rationalize the need for federal involvement by creating a national unemployment fund as a backstop to state funds. In its final report, the committee acknowledged that perhaps the price of progress should be paid in good times as well as bad.[27]

But the bad times arrived sooner than expected. When the stock market crashed in October 1929, Wagner's legislative package emerged from political irrelevance. Suddenly his bills seemed an almost too-tepid response to the gathering economic crisis. With businesses and banks failing and more and more people out of work, the 71st Congress became consumed in a mounting blame game. During a raucous March 3, 1930, floor debate over Wagner's legislation, senators mounted attacks on the Hoover administration's inaction. Senator Hugo Black (D-AL), who would later offer his own legislation to require a shortened workweek, used the debate to urge the enactment of immigration restrictions to avoid further job loss. Meanwhile, Senator Burton Wheeler (D-MT) accused the AFL of spreading rumors about a former Communist candidate for president en route back to the United States from Moscow, allegedly with money provided him for staging major demonstrations of the unemployed in major US cities. La Follette Jr. aligned himself with Wheeler in decrying the use of diversionary red scare tactics. "We cannot solve the problem [of unemployment] by hysteria and the 'red alarm,'" the Wisconsin senator declared. Finally, Senator Carter Glass (D-VA) rose and, with a Solomonic air, suggested that a commission be appointed to study the extent and causes of unemployment. It would not be the last time such an idea was raised.[28]

As Wagner sought to reclaim his time, the unruly debate on his bills threatened to devolve into a series of contentious claims over who was most to blame for the sudden economic collapse. Wagner sought to refocus his colleagues' attention by observing that regardless of who was to blame, the government first needed to determine finally and accurately how many were out of work. Then it could argue over why. As Wagner rightly cautioned, "Much has been said during the last few years concerning technological unemployment. How much do we really know about it? That men are being displaced by machines is of course obvious. But we know nothing of the extent of the displacement." Amid the sudden disorientation caused by economic panic, he thus urged quick approval of his bills so the BLS could begin accurately counting the unemployed on a monthly basis and so that efforts could be made to train and reemploy the jobless.[29]

From Harding's New Normalcy into the Coolidge era, the Republican Party was not doctrinaire in advocating limited government. For example, on policies governing maternity care or aid for wounded veterans, Republican majorities sought an expanded role for the federal government. But even confronted with a growing unemployment crisis, the Senate and House Republican majorities remained steadfastly opposed to Wagner's bills. The majority Republican leadership in the House referred Wagner's bills to an unsympathetic Judiciary Committee. The committee's Republican chairman, George S. Graham (R-PA), told Wagner that although he was sympathetic to the aims of the legislation, he thought much of what was being proposed should be left to the states to execute. When AFL president Green testified, he offered a curt response to Graham's legalistic invocation of states' rights. Green described how unemployment was so acute in the city of Detroit that the unemployed were congregating in city parks, despairing among themselves and desperate to find work. Similar gatherings occurred in other larger cities of the country as the number of those out of work increased. Green thought Graham's reference to states' rights totally missed the mark. "Imagine a cold legal argument of that kind appealing to a man out of work, seeking employment, looking to some source, to some help and finding none," he argued. The AFL president tried to convince the chairman that the conditions leading to unemployment were changing so dramatically that the federal government now had to intervene. Green acknowledged that some states were attempting their own good-faith efforts to help the unemployed, but those states, he observed, were typically unaware of jobs available within their own borders, let alone even in neighboring states. Clearly the states were not collaborating to share information. Wagner's legislation would help alleviate that.[30]

Prior to his testimony, in an article in the *Atlanta Constitution*, Green again warned that neither the creation of new jobs in emerging service industries nor the growth of the professions could overcome the sheer numbers of new workers entering the workforce. He consistently claimed that 9 percent of the working population was idled because of machines. Speaking for organized labor generally, Green increasingly accepted the reality of technology's impact on jobs. He argued that union members simply sought the opportunity to move to new jobs and to afford the products made possible by technological advances. Arguing before the House Judiciary Committee, he noted how the employment exchanges that Wagner proposed were essential to employment equilibrium. If new jobs were opening as industry spokesmen claimed, then workers needed better information as to where they were available. More fundamentally, Green challenged industry to expand and provide even more jobs by making cars affordable to the seven million families who could not afford them. Repeating a theme that would become a staple of organized labor, he reassured the committee that technology was not the enemy. Blue-collar workers simply wanted to share the benefits of increased productivity and to learn about job opportunities in other states, as Wagner's legislation prescribed.[31]

Hoover's Faith in Technology

As Wagner's bills limped forward, President Hoover, the inveterate engineer, held steadfast to his belief that the nation's remarkable advancement was attributable primarily to its superior technology. He embraced the notion that advancing technology was ultimately good for the working man. Just as technology was critical to feed a starving Belgium after World War I, Hoover saw it as integral to economic recovery. His initial response reflected the conventional neoclassical economic view that the crisis could be addressed by bringing production and employment into harmony by balancing supply with demand, and prices with wages. As president he initially had every confidence that those temporarily displaced by machines would soon be reemployed as technology created new opportunities. Though he conceded that in the interim the displaced might need temporary assistance to identify new opportunities and retrain, he was not willing to lend the administration's support to Wagner's legislation that sought to do just that. Only gradually would the new president begin to see technology as a possible cause of escalating unemployment.[32]

When Green wrote Hoover expressing alarm over the way in which machines were reducing the number of jobs, the president, always thirsting for new and better information, acknowledged that he was indeed interested in the impact that new technology was having on job loss. Not only did he tell Green that he might establish a presidential commission to address it, but he also invited Green to become part of an ongoing Presidential Advisory Committee on Unemployment. Hoover conceded that "the very rapidity of our inventions and discoveries has intensified many problems in adjusting what we nowadays call technological unemployment." In seeking answers to the unemployment crisis, Hoover's advisory committee, unlike the earlier conference he chaired for Harding, reflected a balanced membership from labor, industry, and government. It consisted of two subcommittees: the first addressed the broader issues Wagner raised, regarding how to measure the number of unemployed more accurately, and the second was formed specifically to address technological unemployment.[33]

As he initiated this investigatory committee, Hoover responded to the growing concerns over technological unemployment. In a speech the president made in Boston to the AFL on October 6, 1930, he seconded Green's position that firms embracing new technology and thus improving their efficiency should find ways for workers to share in the firms' profitability. Labor should welcome these new machines, the president said, but in doing so, he agreed that employers should share the savings from reduced costs with their employees and customers. On this point, Hoover naturally received boisterous applause from the union audience. For those still clinging to their factory jobs, sharing the gains from increased productivity was consistent with Green's position. At the same time. Hoover used his speech to also express concern for those workers impacted by new factory technology. Revising his administration's earlier estimates, he conceded that the new estimate was that some two million workers alone were displaced by machines. Using the mining industry as an example, an industry he himself had helped modernize, the president noted, "All these conditions [the introduction of machinery to replace men in mines] have culminated in a demoralization of the industry and a depth of human misery in some sections which is wholly out of place in our American system." Yet his empathy did not lead to a call for immediate action. In fact, Hoover's speech also reaffirmed his underlying faith that technology would prove to be the remedy for unemployment. He assured the gathered union leaders that whole new industries created by the next generation of technology would soon be the ultimate antidote to machine-induced job loss. Hoover felt strongly that new innovations would dramatically increase the standard of living and create new jobs in new industries, absorbing those who found themselves out of work. The assurance defied his empirical training as an engineer, since he had no basis for the claim. His own director of the BLS, Ethelbert Stewart, was left to fend for the administration before Senator Couzens's Labor Committee, and Stewart found himself at a loss to tell Congress whether technology in the aggregate was taking or creating more jobs. Moreover, for those whose jobs were being eliminated, Stewart had few answers for Couzens:

> STEWART: This new machine that proposes to set type on the piano-player scheme and operate a linotype by telegraph 500 miles away, 25 or 30 of them at a time, is eventually going to practically end the printing trade.
> COUZENS: Then what will the printers do after all of their training?
> STEWART: I do not know.

Though the president might have believed that technology was creating just as many jobs as it was eliminating, Stewart admitted to Couzens that the evidence, still being gathered selectively and industry by industry, focused on those employed and not those out of work.[34]

Wagner's Initial Triumph

As the Depression moved into its third year and the breadlines steadily grew, Hoover's faith in technology's resuscitative role seemed to defy the bleak reality. Yet the president's penchant for data led him to place Ewan Clague, then an expert statistician at Yale, at the helm of his Presidential Advisory Committee on Unemployment. Clague was asked to decipher whether more jobs were being created by new technologies than were being lost. He began by attempting to redefine the very term "technological unemployment." The committee focused on identifying at least some rough approximation of the rate of increased productivity due to technology and then attempting to see whether there might be a correlation with the rate of employment turnover. In other words, Clague convinced the committee that if one could determine the pace at which industry was adopting new technologies, then perhaps it might allow for a more precise determination of future employment counts. Clague observed that in the two industries where there were some good data on rates of unemployment (i.e., automobiles and rubber), the rapid increase in the use of machines was dramatically advancing output per man-hour. Yet contrary to what most believed, Clague suggested that the data showed that factories employing new machinery were taking on new workers. Though he could not be certain, he speculated that those companies that were not modernizing were more likely losing more employees than those that were not. It was precisely the sort of conclusion the president desired.[35]

In making his case, Clague cited the work of a young University of Chicago economics professor, Paul Douglas, who was an early supporter of Wagner's effort to get an accurate count of the unemployed and to find ways to ensure their reemployment. Yet Douglas was reluctant to make technology the bogeyman. He suggested that any initial displacement of workers due to the introduction of new technologies would not necessarily become a permanent condition; in fact, he believed technology might have positive ramifications for employment. Later, as a US senator and a pillar within the liberal-labor coalition fighting for solutions to unemployment, Douglas would remain less concerned with technology's impacts than many of his fellow reformers. In a view he seemed to share with Clague, Professor Douglas suggested that technologically advantaged firms, with their greater industrial efficiency, would over time become more profitable, expand their operations, and employ more workers. Douglas optimistically declared that "in the long run the improved machinery and greater efficiency of management do not throw workers permanently out of employment nor create permanent technological unemployment." In his report to the advisory committee, Clague built on Douglas's findings by concluding that "mechanization does produce change, and jobs are lost in the process, but in the long run the displaced labor is reabsorbed in the system." Yet his assertion was more speculative than empirical, and his imprecise methodology was reliant on data from essentially just two industries.[36]

Yet Clague's report conveniently reaffirmed the president's basic views about technology and job creation. By acknowledging the still-significant gaps in the information available about unemployment, the report also became a blueprint for an expanded BLS that Clague himself would eventually end up directing for almost two decades. Though generally the report received little press attention, the *Wall Street Journal* suggested that the advisory committee's singular contribution was its reminder that the government still lacked adequate data around unemployment. With the growing Depression at last providing the needed impetus, Wagner's measure to improve the counting finally meandered its way to final passage. On July 7, 1930, Hoover signed the legislation to expand and improve the census of unemployed workers, and Wagner turned his attention to ensuring that additional appropriations were secured for the BLS so that it could widen the scope of its counts and improve their tabulation.[37]

Meanwhile, the senator's bill that required the forecasting of economic downturns and planning for public works spending to ameliorate the impacts of those downturns remained deadlocked in a House-Senate conference committee throughout the summer and fall of 1930. The administration was ramping up its own public works spending as the Depression worsened. Feeling the weight of the growing crisis, Hoover pronounced it his job to find work for what he estimated to be some twelve million unemployed. He called on Congress to authorize $500 million in public works projects. Wagner, sensing that his bill, with its own public works feature, might now garner the administration's interest, reached out to Colonel Arthur Woods, the head of Hoover's Emergency Committee on Employment. Their discussions took a new turn when the Republicans lost both their House and Senate majorities in the 1930 off-year elections. The administration now knew that Wagner would have more political leverage in the next Congress and thus quickly began negotiations. The compromise version (S. 5776) introduced by Wagner on January 5, 1931, retained many of its original planning features and importantly became the precursor for later measures that also sought to smooth the economy by avoiding sudden shifts in unemployment. But after signing the bill into law on February 10, the president largely circumvented its dictates. Hoover's bias for voluntarism and states' rights, coupled with the bill's loose language, allowed him to minimize the role of what was already a neutered version of the planning mechanism in the original bill. The administration adroitly deflected and delayed Wagner's vision of a proactive federal planning capability that might predict future economic downturns by denying it funding. Nonetheless, Wagner's determination that the government needed to plan for future downturns would survive as a central plank of the liberal-labor response to unemployment, and repeated efforts would be made to accomplish what he first proposed.[38]

Meanwhile, Wagner's remaining bill, which established a federal employee exchange, met with continued opposition from a combination of Republicans and southern Democrats clinging still to their states' rights arguments. As usual, however, there were some exceptions in the more progressive Republican ranks. Congressman Fiorello La Guardia (R-NY) challenged the authors of a substitute for Wagner's bill when it finally came up on the House floor. The soon-to-be mayor of New York City noted to his Republican friends that the substitute did nothing to empower the states, and that actually one of the features of Wagner's bill was that it encouraged the federal government to work with existing state employment agencies to create some uniformity in their approach to job information sharing. Meanwhile Hoover's secretary of labor, William Doak, proposed the creation of a new federal agency in every state. Large numbers of Republicans, though still in the majority in the 1931 lame duck session, were nervously looking over their shoulders as the Depression grew worse. Faced with Doak's cumbersome alternative, they suddenly seemed more sympathetic to Wagner's bill and their colleague La Guardia's invitation to vote for it. Though it passed the Congress overwhelmingly, it quickly fell victim to Hoover's pocket veto as the Congress adjourned. Meanwhile the president requested yet another inquiry, this time asking his experts how a "system of placement" might be designed to address technological unemployment.[39]

The Paralysis of Analysis

As Hoover prepared for the presidential campaign of 1932, he asked Fred Croxton to chair the President's Organization on Unemployment Relief (POUR). The panel was to make yet another effort to measure the jobs impact of technology. Croxton, who was serving as head of the administration's Emergency Relief Fund, was administering the disbursement of some $300 million in relief and thus was aware of the mounting unemployment crisis. A Columbia University statistician who later would serve in various capacities in the Franklin D. Roosevelt administration, Croxton had earlier proposed that a permanent government commission he called the National Progress-Prosperity Commission be empowered to regulate the pace at which industry adopted new labor-saving equipment. Croxton's proposed commission was to gradually phase in or stop the installation of any machine that would displace labor. So concerned was Croxton about technology's impact that POUR recommended that the installation of labor-saving equipment should occur only after those workers displaced by any new machine were guaranteed a job with an equal wage. As with Wagner's earlier legislation that created the FESB, Hoover's POUR thus actually recommended that technological advances be forecast and thus calibrated. Yet, just as Hoover's administration worked around the authorization of the FESB, POUR's analysis and recommendation simply went onto the growing pile of such recommendations amassed by the administration's plethora of studies and reports.[40]

As the Hoover administration continued to steep itself in analysis and then often ignore the findings, the human realities of unemployment manifested in social unrest. The so-called "Bonus Army," made up of World War I veterans, many out of work and broken by the Depression, were tired of waiting on a deferred pension. In the summer of 1932, they gathered in the capital seeking early payment of the government bonus promised to them. For Green and his union colleagues, they constituted a surrogate army of the unemployed. "They represent a desperate situation which has arisen out of the continuous unemployment period which has extended over the last three years," Green told his colleagues. The Bonus Army soon found itself confronted by active US soldiers led by General Douglas MacArthur. Though the president did not personally order the attacks of cavalry wielding fixed bayonets, the melee exacerbated an already growing suspicion that Hoover was simply insensitive to those out of work. Though he would leave Roosevelt a stronger infrastructure and reservoir of ideas for economic recovery than was generally understood at the time, Hoover's air of detachment and seemingly insatiable preference for more and better information became an increasingly unpopular political brand as the Depression grew worse.[41]

True to form, just prior to the 1932 election, Hoover asked his Research Committee on Social Trends for yet one more comprehensive review of the causes of the Depression. He thought it would be the blueprint for his second term. Comprising an impressive array of economists, engineers, and management experts, the committee issued its results shortly after FDR's landslide election, which itself served as a staggering referendum on the state of unemployment. "Americans had become painfully alive to the rapid growth of technological unemployment," this final Hoover analysis concluded. Moreover, it acknowledged that the impacts from new machines were more devastating than publicly understood. During the committee's deliberations, the Columbia economics professor Leo Wolman, who would soon serve in Roosevelt's National Recovery Administration and chair the National Labor Board, argued that technological unemployment was now the paramount issue facing the economy. But the experts remained divided over whether the technology that seemed to be creating the problem could also provide the solution. W. F. Ogburn, a sociologist from the University of Chicago, acknowledged that public policy always lagged behind technological innovation. Earlier in the decade, Ogburn had argued that "material-culture changes force changes in other parts of culture such as social organization and customs, but these latter parts of culture do not change as quickly. They lag behind the material-culture changes; hence we are living in a period of maladjustment." Though Ogburn subscribed to the Hooverian faith that the pace of innovation was such that new discoveries would eventually guarantee new jobs for the displaced, the question was when. As the unemployment crisis grew worse, Roosevelt's new labor secretary, Frances Perkins, would begin to question whether public policy could respond as quickly as technology demanded.[42]

On January 31, 1933, during the lame duck session of the 72nd Congress, James Davis, now the first-term senator from Pennsylvania and former secretary of labor under three presidents, rose on the Senate floor to register his disdain for what amounted to a short-lived, though intensely popular, political movement that sought to replace democracy with "technocracy." The iron puddler who had lost his very first job to technology now lent his weight to the mounting critique of the so-called "technocracy movement." As he issued his stinging criticisms of the fast-disintegrating movement, Davis entered into the record the most recent BLS report on the impacts of technology on unemployment. The BLS's extensive industry-by-industry survey was filled with examples of new machines and their contributions to productivity and to technological unemployment. As Davis freely acknowledged, the bureau was clear in its emphasis on jobs displacement resulting from technology.[43]

Since Davis's days as labor secretary, thanks to the steady voices of Senator Wagner and AFL president Green, among others, the plight of the unemployed had gained standing. The Depression now catapulted their cause to a national priority. Reflecting his own blue-collar moorings, Davis warned that the working man was "tired of being frightened" by illusory visions of a future that cast the ordinary worker to the side in favor of a ruling technocratic regime. The technocrats want to solve all of the world's problems, Davis said. It would be nice, he added, if they could solve just this one. From the Harding Conference, to Wagner's insistence on at least counting the unemployed, to the BLS analysis that focused on technological unemployment, in a little over a decade the cause of the unemployed that once struggled for attention now became the singular focus of a new administration. The new Roosevelt administration and the growing liberal-labor coalition that supported its New Deal were asked to respond with solutions as innovative as the technology that was displacing so many workers.[44]

CHAPTER 2

Taming Technology

> *Industry has exploited labor through the medium of machines until labor is placed on the scrap heap, with no purchasing or consuming power.*
>
> —Letter from Edgar Boles, president of Local Union 106, Harrisburg, IL, to the House Labor Committee, February 17, 1936

On New Year's Day 1933, the Republican-leaning *Chicago Daily Tribune* greeted readers with the banner headline "Here's What Ails the U.S.A." The economy was reported to be growing worse by the day. The BLS estimated that 25 percent of the workforce was now out of a job. US workers feared for their economic future, the article reported, and they yearned for Congress and the new president to quickly devise more innovative policies to meet the crisis. An accompanying article, however, indicated perhaps a reason for cautious optimism. The president-elect, still two months away from assuming office, was reported to be busily meeting with civil servants inside agencies and departments. In effect, he seemed to be scouring the government in search of any new insights or ideas on how best to address the growing economic crisis. FDR was also reported to have invited congressional leaders to meet with him and his "brain trust" at his transition offices in Hyde Park, New York. Though the *Tribune*'s front-page article was based mostly on rumors and secondhand reports, those in contact with the president-elect's transition operations sensed the optimistic energy that engulfed the transition.[1]

With FDR's inauguration on March 4, 1933, the causes of unemployment suddenly became far less important than the cures. In the administration's fabled first hundred days, Roosevelt and his advisers wasted no time in implementing a dizzying array of new ideas to address the failing economy.

The new administration dove headlong into the business of job creation. In addition to pursuing their New Deal agenda, they doubled down on many of the initiatives already taken by the outgoing administration. They also gave new political life to legislation promoted by longtime champions of the unemployed, such as Senator Wagner. Roosevelt, with his effusive confidence and sunny disposition, quickly captured the public's confidence. Workers who typically shouldered personal accountability and blame saw in the new president a sympathetic ear. They wrote to him to share their sense of having a personal responsibility to find a job. In his first week in office, almost half a million letters came to the White House. FDR's personalized style provided citizens with a new outlet for their concerns, and more importantly a reason for hope. "I have never as yet begged, but must and will be very candid, that I would appreciate some kind of help," wrote one citizen to his new president. Often these letters asked FDR where they might turn for assistance in the suddenly growing governmental apparatus aimed at helping them.[2]

These letters reflected how ingrained the work ethic was in the American ethos. In their energetic response to the unemployment crisis, FDR and the emerging liberal-labor coalition that supported his New Deal would soon go well beyond Wagner's three groundbreaking bills to enact a remarkable array of measures that put people back to work and compensate those employees who lost their jobs. Wagner, at the core of the liberal-labor coalition, would see through proposals that enabled blue-collar employees to exact their fair share of the gains from productivity increases that continued despite the larger economic malaise. And yet, as it achieved its impressive arsenal of reforms, Roosevelt's New Deal seemed unable to stem the persistently high rates of unemployment. During a 1935 press conference in Hyde Park, FDR would echo his brain trust's increasing belief that technology was having a more systemic impact on employment than he had earlier cared to admit. While the precise role the new machines were playing remained difficult to measure, technology would now receive its share of the blame as the economy sputtered. By decade's end, even a new economic philosophy that encouraged using fiscal policy to spur demand seemed unable to resuscitate a still-ailing economy, with its disturbingly high rates of unemployment.

Finding and Keeping Work

From the very outset of Roosevelt's energetic administration, the president singularly focused on how to help the unemployed worker get back on the

payroll. Having seen the value of Tammany Hall relief programs in building political loyalties, he sought to adopt the same approach by federalizing
unemployment relief efforts beyond what Hoover had already begun with
the Reconstruction Finance Corporation. More than bolstering meager-to-
nonexistent state aid programs, Roosevelt wanted to immediately create
new jobs. And, unlike his analysis-obsessed predecessor, Roosevelt mapped
out large problem-solving visions and then allowed his staff to tend to the
mechanics of implementation. In addition to creating the Civilian Conser-
vation Corps, among the first bills FDR signed was the last of Wagner's
original bills, which proposed creating a federal employment service to
coordinate job information sharing among the states. For all the struggles
Wagner's bill had earlier endured, this remaining measure of the three he
had first introduced in the 70th Congress sailed through the 73rd Congress
in the first three months and was signed into law by the president on June 6,
1933. When the bill was debated in the House, a strapping young congressman from the panhandle of Texas, soon to catch the eye of the new president, asked the bill's House floor manager, "Is this measure substantially the
same as the bill known as the 'Wagner bill' which was vetoed by President
Hoover?"

"It is," replied the House Labor Committee chairman, William P. Connery
(D-MA). That was all the answer Lyndon Baines Johnson and others in
Roosevelt's New Deal coalition needed to pass legislation that would allow
the unemployed to learn of available jobs.[3]

Complementing Wagner's job placement legislation was the president's
Federal Emergency Relief Act (FERA), passed the month before. It was pat-
terned on a New York state program Roosevelt had put in place as governor,
and FDR decided to ask the head of the Albany program to operate this
largely identical national program. Harry Hopkins was a product of rural
Iowa, a descendant of the social gospel movement, and well networked with
charity and relief efforts across the country. In the first few hours after taking
the position, Hopkins disbursed some $5 million to the states, and then some
$500 million more over the course of FERA's first year. Roosevelt empha-
sized that the legislation was meant to challenge the states to organize their
own relief efforts. Predictably this charge met with uneven results. Hopkins
was intent on the states professionalizing their own welfare offices and was
revolted when some states disbursed the relief in cash handouts. "Three or
four million heads of families don't turn into tramps and cheats overnight,"
Hopkins lamented as he pushed back on lax state administrators who seemed
more intent on delivering cash than on creating work relief. As the administration debated the states over how to dispense aid to the unemployed,

Wagner and his progressive allies decided that the flawed state response to FERA demanded a new federal system with more exacting requirements. His proposed federal unemployment compensation proposal would soon become an important feature of the Social Security Act.[4]

As the severity of the Depression grew, and even as Roosevelt's new programs offered some rays of hope, many anxiously clung to their jobs, knowing that their days of employment might be numbered. Initially the AFL's Green was careful not to endorse blanket federal welfare legislation. He hoped labor would be able to continue to exact concessions from management by bargaining rather than legislating. This position by organized labor effectively left the unemployed depending on the new administration and its congressional New Deal coalition for assistance. But over the course of the Great Depression, as unemployment became the new norm, union members themselves joined their already out-of-work cohorts in vocally expressing anxiety about how new invasive machines could soon eliminate their jobs as well. In Indiana a union sought the physical removal of machines that were taking away mining jobs. Members of the rail unions asked their leadership to support legislation that would prohibit the introduction of any new machines in the workplace altogether. As this worker resistance to technology grew, union leadership sought new ideas that might channel the growing antimachine animus among its rank and file. A new "Share the Work" legislative proposal provided labor just the opportunity it sought to slow the pace of technological development and thus ease the rate of unemployment.[5]

Beginning in February 1933, Senator Hugo Black (D-AL), reelected in November for a second term, seized on the momentum created by the new "Share the Work" movement that began at Ford and spread to other firms (e.g., Kellogg's and Goodyear). Those companies found that reducing the number of hours each employee worked minimized the number of workers subject to layoffs. Seizing on the idea, Black convened hearings in his Judiciary subcommittee on his proposed legislation that mandated a six-hour day and thirty-hour workweek for any firm engaged in interstate commerce. It seemed like only yesterday that William Green was urging an eight-hour day and a five-day workweek, and Black's proposal was a stark indication of how dramatically economic conditions, and the union position along with them, had turned. Witness after witness testified before Black's subcommittee as to how productivity increases were eliminating jobs. It was reported that since 1928, technology had increased productivity in the tire industry by 86 percent. Railroad representatives testified that from 1919 to 1929, their technologically more efficient operations combined to cause a loss of some

350,000 jobs. A textile mill owner told the subcommittee that because of significant improvements in advanced equipment, he laid off scores of workers who would not be replaced. Oil and coal operators spoke to the remarkable improvements in drilling and mining equipment that spurred similar productivity gains and job losses in their industries. A representative of the United Mine Workers suggested that the average tonnage mined per man per day had doubled since the turn of the century. And on it went; the litany of examples of men displaced by machines seemed endless, and Black gladly took it all as a reinforcement of the need for his legislation.[6]

Black believed a shorter workweek was just the sort of immediate and direct response that could help alleviate the growing crisis. Among those supporting Black's legislation, Green testified that industry was now so technologically advanced that it was unrealistic to expect that it could any longer provide a traditional workweek. In Green's view, organized labor needed to accept the idea of what he called "a permanent standing army of unemployed" while working to find new ways to prevent this army from finding new recruits. Black's thirty-hour proposal was an option he heartily endorsed. Among the business witnesses from the larger Share the Work movement, Kellogg's president Lewis J. Brown testified that the firm had raised the minimum daily wage for its workers to preempt any worker concerns that a shortened workweek was simply a back door to excuse lower wages. Brown argued that if structured properly, a shortened workweek could give workers the same purchasing power that they enjoyed with a longer workweek. L. C. Walker, who served as the vice chairman of the movement, asserted that "expanding the number of shifts and raising wages would increase the government's tax receipts and reduce the costs that the unemployed placed on the community." With business and union backing, Black and his supporters hoped the idea would garner the endorsement of the popular new president.[7]

Yet despite what seemed widespread support from labor and industry, the administration's own view on Black's bill remained unclear. By the spring of 1933, as the crowded New Deal legislative agenda progressed through both chambers of Congress, Black's bill finally garnered Senate floor time. Senator Joe Robinson (D-AR), the majority leader, offered an amendment to increase the thirty-hour week to thirty-six. Many suspected that Robinson's amendment had the administration's tacit approval. Nevertheless, it was defeated, and the administration remained mysteriously silent on Black's bill. It would soon become apparent that while Roosevelt had some reservations about the practicality of the bill, his real concern was that legislating a reduced workweek would distract political attention from his soon-to-be-introduced New Deal centerpiece, the National Industrial Recovery Administration (NIRA). Roosevelt believed he could accommodate the Share the Work movement by offering shortened work hours under the new and sweeping NIRA authority. Though Black's bill passed the Senate by a vote of 53–30, the introduction of NIRA doomed it in the House, where the administration proceeded to artfully extinguish its chances.[8]

Though Black's bill soon became ensnarled in Congressional procedure, the idea for a shortened workweek remained highly popular among union leaders. They were already suspicious of business influence in the administration's drafting of NIRA legislation. Their general view was that the proposed industrial boards the legislation established gave business far too much authority over wages and prices. Green warned that under the industry-friendly NIRA, a minimum wage would become a maximum wage, and he noted that the unions opposed the bureaucratic procedures detailed in the bill's provisions to reduce working hours. The fissures between labor and the administration over Black's bill reflected larger fault lines that soon emerged between organized labor and the administration.[9]

Though workers and the unemployed were at the heart of the New Deal liberal-labor coalition, organized labor found its own authority and political power at a low point during Roosevelt's first term. Influenced in part by his labor secretary, Frances Perkins, Roosevelt had a penchant for enacting laws that empowered workers rather than unions. Though the president was willing to sign the Norris-LaGuardia Act, which limited the authority of the government to curtail strikes, he constantly battled union bosses on the passage and implementation of NIRA. During the debate over this centerpiece of New Deal legislation, the burgeoning population of unskilled workers, largely represented by the much-caricatured president of the United Mine Workers, John L. Lewis, began to establish its own political identity. Soon Lewis would be instrumental in forming the Congress for Industrial Organizations (CIO) and taking the unskilled worker out from under the umbrella of the skilled-labor-dominated AFL.[10]

Lewis knew firsthand the ravaging impacts of technology on jobs in the coal fields of Appalachia. A powerful, theatrical presence with a bombastic speaking style, he was a shrewd political presence who seized on Roosevelt's generalized empathy toward the working man and used it to grow the power and influence of the unskilled worker in an enlarged and empowered union movement. Lewis forcefully took issue with what appeared to be top-heavy industry influence in the NIRA legislation. Frustrated by its reliance on the establishment of industry codes and pricing mechanisms and railing against what might happen to wages and jobs, Lewis exacted his own price from the administration. In legislation that contained something for virtually every constituency, Section 7a of NIRA established the right of collective bargaining. This singular provision became the crown jewel of labor's Depression-era platform. Though NIRA in its entirety would be overturned as unconstitutional by the Supreme Court just short of two years after its enactment, the initial achievement of unions' collective right to bargain galvanized unskilled laborers to mobilize politically. Upon NIRA's enactment, Lewis famously remarked, "President Roosevelt wants you to join the union!" In the short span from June to October 1933, following NIRA's enactment, 1.5 million workers would join the labor movement. Nonetheless, even at their height during the Depression, unions comprised only one in three workers.[11]

Calibrating Expectations

With the enactment of NIRA, the New Deal was at full throttle. Its already impressive panoply of programs and projects reflected Roosevelt's basic neoclassical economic view, which initially steered him to seek adjustments to wages, prices, and production as the best way to jump-start economic growth. His brain trust emphasized that by regulating these three essential components of the flailing economy, they could restore a healthier balance between supply and demand. The new industry codes were designed to achieve the amount of price and wage adjustment required to modulate demand while setting forth production schedules to regulate supply. FDR's economic engineering was aimed at stimulating consumer demand to bring back into balance what his brain trust perceived as a continued tendency toward overproduction. Indeed, the administration pursued policies founded on the basic premise that adjusted production schedules, stabilized prices, and rising wages would lead to increased consumer demand and new job creation.[12]

As for technological unemployment, many advising the president quietly registered their concern that it would continue to undercut FDR's many New Deal initiatives. Prior to the election, one of FDR's closest advisers, Rex Tugwell, suggested that "occupational obsolescence" was long a part of the industrial landscape. What was increasingly accepted was the idea that gains in efficiency should be more widely shared. Tugwell was part of the Columbia University Economics Department, so formidable in shaping

New Deal policy. Many of Tugwell's colleagues seconded his concerns about the deleterious impact of technology on employment. Tugwell specifically targeted Chicago University professor Paul Douglas's dismissive assertion that technological unemployment was tantamount to frictional unemployment. Tugwell took the opposite view. Technological unemployment, he argued, was not some temporary phenomenon. Rather, the impact of advancing machinery on the workplace was at the heart of the unemployment problem and thus far more insidious. "There is no cure for that obsolescence which is traceable to technological change and to more efficient arrangements," Tugwell asserted. Technological unemployment was a permanent fixture of the industrial economy, and those technologically displaced deserved a much more elaborate social safety net to protect them against its effects. At this time, Tugwell was serving as Roosevelt's agricultural adjustment assistance administrator, and his immediate challenge was to overcome the mechanization of farms with forward-looking plans to help farmers and farmworkers displaced by that mechanization. He extrapolated from what was happening on the farm to sound a cautionary note for the industrial economy as well.[13]

Others whom Roosevelt relied on were similarly concerned about technological unemployment. In her 1934 book *People at Work*, Frances Perkins decried the unemployment situation FDR inherited, one she believed was exacerbated by technology. She viewed the increasing introduction of labor-saving machines as the major hindrance to economic recovery. The key going forward, she asserted, was to learn how to better forecast how the introduction of new machines would impact unemployment in specific sectors and thus be better able to anticipate and plan for the consequences. She called for some measure to plan and anticipate, an idea not all that dissimilar from Senator Wagner's proposed FESB. Perkins said that she wished economists were as gifted in predicting the future as the inventors were in creating it. "If the economists were under the same realistic pressure as the inventor is and had to put their conceptions to the test of reality, we might get sounder and more active thinking in this field," she noted. Perhaps with better forecasting, the administration's plans to increase purchasing power and production rates could be more perfectly synchronized.[14]

Those inside the administration were joined by influential voices outside the administration who also drew attention to the phenomenon of technological unemployment. The economist Stuart Chase's 1932 book *A New Deal* was already credited with shaping many of FDR's initial policies and its brand name. In a 1934 article in the *Washington Post* entitled "Machines

Winning in the Battle for Jobs," he warned that the steady improvement in technology would make it difficult, if not impossible, to achieve former employment levels. Chase confirmed that productivity levels dramatically increased throughout the 1920s and were continuing to do so even when industry might seem starved for capital. According to one Columbia University economist Chase cited, productivity had improved 25 percent since 1929. Noting that in 1932 it took only seventy-six men to do what one hundred men could do a decade earlier, he conceded that even returning to previous boom-year levels of output, the nation would still have unacceptably high levels of unemployment going forward. As an early formulator of much of the thinking that underlay New Deal policies, Chase now urged consideration of a shorter workweek as yet one more attempt to work around technology's corrosive effect on employment.[15]

As the dire warnings about technology's impact grew, the economy seemed to take two steps backward for every step forward. Whether it was coal miners in Appalachia or longshoreman at the Port of San Francisco, increases in their output as measured by the Federal Reserve were not being matched by increased employment. Showing some signs of recovery, the gross national product (GNP) grew 16.8 percent in 1934 and 11.2 percent in 1935. Employment numbers, however, did not seem to keep pace. Unemployment was not declining as precipitously as GNP was rising. By May 1935, news accounts revealed that the unemployment rate was creeping up again. As reported in the *Baltimore Sun*, the BLS was doing its best to estimate jobless numbers based on statistical samples from employment numbers, but independent counts confounded the big picture. One report by the International Labor Organization showed over half a million more unemployed in the United States than in the year before. The Industrial Conference, which did its own survey, noticed a 1 percent increase in unemployment. Despite Wagner's earlier legislation, the official numbers remained frustratingly inexact. Yet it was clear that employment conditions were not keeping pace with the economy.[16]

At this juncture, with little sign of a robust recovery, Wagner now took the lead of the dramatically expanded liberal-labor coalition that formed around New Deal policies to expand the scope of reforms aimed at greater economic security for those both with and without jobs. He offered his own legislation to federalize unemployment compensation by requiring that employers contribute to state unemployment compensation funds. "It is no answer to millions of idle workers and their suffering families to say that within 10 to 15 years something might show up," Wagner warned his colleagues. He believed that FERA, though well intentioned and delivering needed jobs, was still leaving too many in breadlines or on the curbside selling apples. Appearing before a subcommittee of the House Ways and Means Committee in March 1934, Wagner noted that even in the boom year of 1927, between two and three million people could not find work. Now, echoing his friend Frances Perkins, he noted that it was evident more than ever that "the displacement of men by technological improvements suggests a long-range challenge to our social inventiveness." Policy needed to become more creative if it was to thwart technology's ill effects. From his earlier engagement on behalf of the unemployed, and particularly in the depressed economy, he believed the public understood that unemployment was not the result of personal flaws. And yet Wagner also understood how unemployment could take a mental toll. Affirming the still-strong belief in the dignity of work, Wagner declared, "So long as we live under a system in which industrious men normally win their bread by working, industrious men suffer degradation when they must exist by begging." In an industrial society where technology was constantly changing employment patterns, relying on the ad hoc beneficence of charities was inadequate. A permanent system of relief needed to supplement "voluntary systems of protection" that were woefully inadequate for a modern economy.[17]

When Perkins testified on the pending Economic Security Act (soon to be named the Social Security Act of 1935), she told the Ways and Means Committee that they would be surprised to learn the sheer number of new labor-saving devices introduced since the Depression began. Continued technological advancement was a significant factor in keeping unemployment levels well above what the recovering economy deserved. Perkins joined Wagner in arguing that the federal government simply could not create enough jobs on a sustaining basis to address the relentless impacts of technology. It seemed that the awareness of technology's impact on thwarting the New Deal's aims was slowly sinking in with the still-new administration. The government needed to find ways to reach those on the now-too-familiar breadlines with improved assistance, and Perkins and Wagner were in lockstep regarding the need for a federalized unemployment compensation system.[18]

Wagner immediately sought to reduce political resistance from the states concerning the mandatory establishment of state unemployment compensation funds by ensuring that the federal government would maintain the level of contributions needed to keep them viable. Wagner's bill thus created an employer payroll tax that could help reduce the burden on state treasuries. Then, in turn, to reduce industry resistance to the tax, the bill allowed firms to deduct from their federal taxes most of the amount they would pay to the state funds. Wagner's proposal borrowed from the model established in the state of Wisconsin, as designed by the economist John Commons and featured at Couzens's 1928 Senate Labor Committee hearings. Based on the Wisconsin model, the bill provided the federal impetus for other states to follow Wisconsin's lead by lessening the financial burden on both states and corporations. On August 9, the Senate passed the Social Security Act of 1935, which incorporated Wagner's unemployment compensation provisions. Two weeks later, FDR signed the act, pointing out how the bill "gives at least some protection to thirty millions of our citizens who will reap direct benefits through unemployment compensation, through old-age pensions, and through increased services for the protection of children and the prevention of ill health." Two years after the law's enactment, all forty-eight states had adopted some form of unemployment compensation, creating at last a modicum of security for those displaced from work.[19]

While Wagner steadfastly pursued a federalized unemployment compensation system, he simultaneously sought to ensure that existing workers received their fair share from increasing productivity gains. He was concerned that even before the NIRA collective bargaining provisions went into effect, employers were devising their own clever schemes to circumvent its dictates. Indeed, companies were forming their own friendly company unions, enlisting sympathetic workers and asking them to cajole their unsuspecting colleagues. As such evasive corporate tactics spread, Wagner argued that any claim that NIRA was protecting labor rights was "a sham and a delusion," a political dodge that should be prohibited. Thus, on February 21, 1935, with the full backing of organized labor, he offered his own stand-alone legislation to more securely codify labor's right to bargain. The administration, hardly in a position to acknowledge NIRA's shortcomings, was hampered in its support of Wagner's proposal. In fact, the very case challenging the entire NIRA (*A. L. A. Schechter Poultry Corp. v. United States*) was now headed toward the Supreme Court. Yet, thanks to Wagner's early action, on May 1, two weeks before the Supreme Court rendered its own verdict ruling NIRA unconstitutional, the Senate approved Wagner's bill that established the unions' right to bargain. Wagner's National Labor Relations Act (soon to be known as the Wagner Act), now with a legislative head start, took only a few weeks to pass the House. Despite the administration's concerns over the independence of the act's new National Labor Relations Board, in the wake of NIRA being ruled unconstitutional, FDR had little choice but to sign the Wagner Act into law over the July 4 weekend. Wagner's prescience was again rewarded and the union's right to bargain was preserved.[20]

Throughout the debate over Wagner's landmark legislation, the issue of technological unemployment figured prominently. Commenting on what would become one of his signature legislative accomplishments, the senator noted that although it was true that technology was lightening the burden on workers, it was doing so at a dramatic cost. The benefits from increasing productivity were not being adequately shared. Suggesting an alternative view, Wagner noted, "We apologetically referred to technological unemployment when in truth we were suffering from refusal to confer the benefits of technology upon workers as well as owners." It was a reminder of what Green and others in the labor movement had consistently observed over a decade before. Since it seemed impossible that governmental policies could effectively rein in technological progress, workers needed to share in its benefits. With the Wagner Act in place, at least those still employed firmly secured their bargaining rights. Unions would now have an opportunity to exact their fair share of the increased profitability made possible by machines, even as jobs were being eliminated.[21]

On August 5, 1935, with the NLRA now law and Social Security two weeks away from his signature, Roosevelt gathered reporters around him at his Hyde Park family retreat and began fielding a range of questions about staff disputes in the White House and how the remaining portion of a $4 billion Works Progress Administration (WPA) allocation might be disbursed. On this warm summer's day, even FDR's charm could no longer divert the attention of a gaggle of skeptical reporters, who were curious about what new ideas the president might have for reversing the fortunes of a still-sputtering economy. The president, who typically resisted casting blame, decided to use the opportunity to address the nature of unemployment. He hypothetically offered that even if the country's economic machine were to be suddenly restored to the capacity that existed prior to the 1929 crash, it would require only 80 percent of the manpower. He seemed to fully accept the argument that the continued introduction of new machinery was constantly taking jobs. Seizing on the president's acknowledgment, Bernard Kilgore, who would later become managing editor of the *Wall Street Journal*, noted how Roosevelt and his aides seemed to have quietly reconciled themselves to the fact that technology's improvements created a class of the permanently unemployed. The *Christian Science Monitor* reinforced Roosevelt's pessimism, noting how productivity increases were not shared with workers or consumers. "Alas! There isn't a Henry Ford in every industry," its report lamented. FDR now seemed convinced that technological advancement was a corrosive force on employment. Unaccustomed as he was to placing blame, Roosevelt, perhaps eyeing the election season ahead, seemed to be calibrating public expectations as to what more his New Deal might deliver.[22]

Taxing Machines

Entering the 1936 election year, Roosevelt sought to put the best face on a still-wobbly economy. The economic indicators of renewed growth and job creation continued to fluctuate. The AFL suggested that approximately 12.5 million remained out of work, while the Conference Board placed it at approximately 9.8 million. At 9.03 million, the BLS's official number was even lower. The unemployment rate was also down significantly from when FDR took office. Based on subsequent analysis, we know that in April 1933, a month after he assumed office, the unemployment rate reached a high of 25.4 percent. In April 1936, when the president spoke to New York's Democratic Party at a Jefferson Day assembly at the Commodore Hotel, it stood at 15.5 percent. It would trend downward for another year. Yet although the BLS's methods for counting the number of unemployed was much improved since the enactment of Wagner's original "counting" legislation, the process remained imprecise and the final numbers subject to critique. And even though the numbers may have shown some continuing alleviation of the unemployment crisis, Roosevelt's White House continued to receive letters from across the country reminding the administration that whatever the unemployment numbers showed, technology continued to take its toll. One letter read, "Dear Mr. Roosevelt: Please, for the sake of humanity and our Nation, retire enough of the non-consuming, productive, labor-displacing machinery to allow man, the consumer of the products of the farm and factory, to have employment and purchasing power." Though Roosevelt was now attuned to the impact of technology on the persistently high unemployment levels, in his address to the party faithful he exuded a tough-minded, optimistic response. He reminded his home-state friends how his administration ended the long breadlines in the garment district by putting money in people's pockets and breaking the throttlehold of underconsumption. But FDR also expressed his newfound concerns about technology and warned that the constant introduction of new machinery might soon lead to new spikes in unemployment.[23]

As the Depression wore on, some in Congress also began to reexamine the relationship between technology and employment. House Labor Committee chairman John Lesinski's (D-MI) Dearborn, Michigan, district was populated with an increasing number of idled autoworkers. On February 1,

1936, Lesinski gaveled in the first of five days of hearings on a cumbersome resolution that called for the Labor Department to complete a comprehensive inventory of the labor-saving devices in operation since 1912, then develop estimates of how many of the unemployed could be attributed to the adoption of each device, and finally project how many workers would have been employed if these machines were never introduced in the first place. The measure, in its sheer complexity, if not impracticality, seemed more symbol than substance. Introduced by Baltimore congressman Vincent Palmisano (D-MD), who separately chaired the House Education Committee, it nonetheless was a reminder of how policymakers remained challenged by the inability to forecast technological progress.[24]

Palmisano applauded Lesinski's call for an analysis of the rate of technological adoption as the correct pretext for imposing a tax on the introduction of new machines in the workplace. Only then, he and a growing number of supporters argued, could the nation slow the pace of technological unemployment. Congressman John Hoeppel (D-CA) offered a "machine tax" measure in the previous Congress, and the Maryland congressman hoped that by adopting his resolution, Congress would begin to build the substantive record necessary to advance the idea. A machine tax possessed the political virtue of being straightforward and simple. Though the grassroots movement that rose to support it lacked a national leader, the idea was enthusiastically promoted in articles and pamphlets produced by newly formed organizations such as the American Techno Tax Society in California and the National Organization for Taxation of Labor-Saving Devices. Their efforts produced a steady dose of letters to the White House urging adoption. During his testimony, Palmisano made his case for how such a tax would slow the pace of technology's advance in the workplace. After all, he argued, if the federal government could grant corporations legal authority in the first place and protect their patents, then why shouldn't it be able to levy a tax on machines to save more jobs? "If you find that the machinery is supplanting the men and they are unable to earn a livelihood, then slack up and permit them to work," Palmisano urged the committee. Once it was ascertained how many men were unemployed because of a machine, as his pending resolution proposed, a prorated tax might be enacted, targeting those companies that produced or installed them. The congressman emotionally detailed by name those businesses in his own Baltimore neighborhood that laid off employees or went out of business because they could not keep pace with the "monopolists" that could afford to invest in the latest job-replacing technology. Perhaps a tax would have made their owners think twice. Palmisano's heartfelt testimony before Lesinski's committee captured the chairman's attention. He declared, "There is only one thing I can see to be done and that is to tax these machines."[25]

Though some of his own analysts, reflecting the suspicions of labor activism, dismissed the tax solution, FDR did little to discourage the idea that something needed to be done. Ever since his Hyde Park press conference, the president had continued to share his own concerns about technology. FDR reflected what now seemed increasing public anxiety in a letter he wrote to a leading public relations adviser: "I suppose that all scientific progress is, in the long run, beneficial, yet the very speed and efficiency of scientific progress in industry has created present evils, chief among which is that of unemployment." And while the idea for a so-called techno-tax never gained legislative traction, the idea continued to resonate in union halls and among public intellectuals. With the November elections fast approaching, though FDR may not have supported the machine tax, he asked his trusted adviser Harry Hopkins to look into the causes of unemployment, with particular emphasis on the role that technology played. Having spent his first term seeing through a plethora of reforms and programs to spur job creation, the president was concerned that the recovery remained elusive.[26]

Hopkins asked Corrington Gill, a skilled government statistician in the WPA, to help coordinate a study that would place heavy emphasis on determining the effect of increased mechanization on employment. The mega-project would last until 1941 and result in the release of over sixty separate reports. Writing in the *Los Angeles Times*, Gill suggested that the comprehensive WPA review would explain the impacts of technology on jobs once and for all. In Gill's view, those who saw technology as the solution to the nation's unemployment woes were the mirror reflection of those who saw it as the principal barrier to recovery. Both sides accepted that labor was displaced by machines. Both sides also agreed that labor was being absorbed by new industries, including those servicing those industries. Where the disagreement lay was over how extended the lag was between separation and absorption. "One group minimizes or ignores the lag; the other emphasizes or exaggerates it," Gill observed. He in turn assigned David Weintraub the task of determining which side was closer to the truth.[27]

Weintraub's initial analysis, *Unemployment and Increasing Productivity*, concluded that "unless substantial increases in production accompany the technological advances and population increases of the last eight years, the nation will be confronted for some time to come with serious problems of industrial, economic, and social readjustment." With the Depression entering its eighth year, the New Deal seemed to be reaching a point of diminishing returns. The report acknowledged, "There are no signs of any brakes on constant accelerated technological advance," and concluded that the dramatic increases in productivity over the previous decade were a root cause of the sluggish employment numbers. The president's instincts were correct. From 1920 to 1929, the national output (i.e., GNP) increased some 46 percent, yet with only an accompanying 16 percent growth in the labor force. From 1932 to 1935, production volumes again increased, but the rate of employment again substantially lagged. Echoing what Roosevelt hinted at during his Hyde Park press conference, Weintraub determined that the country needed to produce 120 percent more than it did in 1929 to attain the same low unemployment rates it had experienced before the market's crash. Weintraub's report thus confidently concluded that technology was having economy-wide impacts spreading well beyond any one factory site or single industry. In short, he wrote, technology was responsible for "a considerable proportion of the unemployed."[28]

Fair Labor Standards

Yet despite the gloomy assessment and faltering economy, Roosevelt's landslide victory in the 1936 election was a testament to how his many New Deal initiatives cemented the unwieldy constituencies that made up the liberal-labor coalition. Its strong urban moorings drew support from the increasingly populous ranks of immigrants, African Americans, and unionized blue-collar workers, yet also from southern farmers and the growing number of poor white southerners benefiting from the 1933 Tennessee Valley Authority Act. Roosevelt's farm relief measures, which helped blunt the impacts of the mechanized farm, were so popular throughout the rural United States that Republican candidate Alf Landon was unable to hold even his home state of Kansas. Roosevelt entered his second term with a stronger mandate than he'd had during his first term. He won all but eight electoral votes and his long coattails strengthened Democratic majorities in both the Senate and the House. Undoubtedly Roosevelt's electoral success was also aided by what seemed a declining unemployment rate, which, though incrementally improving, still hovered at around 15 percent. Six months after the election, however, the slow if unsteady recovery stopped and the nation suddenly fell back into what was termed the 1937–38 recession within the Depression.[29]

With Democratic majorities now firmly in place in the 75th Congress, and the short-lived economic recovery fading, FDR went right to work attempting to reenergize his New Deal liberal-labor coalition in support of the Fair Labor Standards Act (FLSA), which would establish a minimum wage and a standard forty-hour workweek. Ironically, Senator Black, whose thirty-hour-a-week proposal Roosevelt earlier sidelined, now chaired the Senate Labor Committee, which conveniently held jurisdictional claim to the legislation's future. During the FLSA's thirteen-month legislative odyssey, Black and other New Deal southern progressives, like the rising star Lyndon Johnson, assumed a new significance in the administration's political calculus. In the aftermath of the southern Democrats' filibuster of antilynching legislation, Roosevelt understood how the indefatigable Black would be instrumental in cajoling and compromising with his fellow southern Democrats. Roosevelt also had to persuade union leaders that it was important to legislate rather than simply negotiate minimum wages and hours. Union leadership continued to believe that wage and hourly concessions were best achieved in negotiations with management, not legislators. Nonetheless, though FDR felt confident he could persuade labor to come around and support the FLSA just as he had with earlier New Deal proposals, the conservative southern Democrats posed a far more significant challenge. Attorney General Homer Cummings noted that when the subject of a minimum wage was mentioned, many southern Senators "actually froth at the mouth." Many of Black's southern colleagues were now seizing on depressed economic conditions to actively court companies in the industrial North by touting the South's inexpensive labor. A federal minimum wage undercut the potential for lower wage locations in the South, with its substantial unskilled, postagrarian worker population. Predictably, then, the minimum wage that Black now proposed on behalf of the administration split southern Democrats. Their resistance proved formidable and, coupled with organized labor's tepid support, frayed the New Deal coalition at the peak of its power. By the time the bill made its way through the process, it was scarcely recognizable.[30]

Throughout the FLSA's tortuous path as it faced opposition from an emerging coalition of midwestern Republicans and southern Democrats, the administration and its legislative champions—including its lead Senate sponsor, none other than Robert Wagner—repeatedly turned to the growing specter of technological unemployment as justification for expediting the bill's passage. New Deal lieutenants such as Massachusetts Democrat Arthur D. Healey (D-MA) were quick to note that, unlike the Republicans, the Democrats promised "to do something about the vast problem of technological unemployment." In support of a bill that promised "a fair day's pay for a fair day's work," Senator Healey promised the elimination of the worst effects of machines on jobs. In the House, Congressman Henry Ellenbogen (D-PA) rose to remind his colleagues that the only way to eliminate unemployment caused by new machines was to give workers a decent minimum wage: "It is said by leading economists that if we could raise the purchasing power of the underprivileged one third of our population, we could obtain such an increase in our purchasing power that would enable the operation of all of our factories at full capacity." It was a sweeping claim and only vaguely supported. Yet it again embraced the liberal-labor notion that workers should share in the gains from industrial progress. On June 24, 1938, as he prepared to sign the FLSA, FDR told his fireside chat listeners that "without question it [the FLSA] starts us toward a better standard of living and increases purchasing power to buy the products of farm and factory."[31]

The Great Depression was now in its ninth year, and as the summer wore on, the economy descended to greater depths. Compounding the unemployment conundrum, the latest iteration of the extensive Weintraub analysis showed that older workers, blacks, and youths entering the labor market faced particularly steep challenges in finding scarce jobs. Many had not held jobs for five years. Nonetheless, FDR seemed to always see the glass as half full. As one biographer noted, "His inveterate optimism gave him the wherewithal to see beyond the dark moments so demoralizing to many others." With midterm elections approaching, he was quick to point out the hypocrisy of his business opponents. In a fireside chat, he told Americans, "Do not let any calamity-howling executive with an income of $1,000 a day, who has been turning his employees over to the Government relief rolls in order to preserve his company's undistributed reserves, tell you—using his stockholders' money to pay the postage for his personal opinions—that a wage of $11 a week is going to have a disastrous effect on all American industry." Though unemployment was reported to be near 20 percent and recovery nowhere in sight, FDR proudly signed the FLSA, hoping that raising the minimum wage and increasing overtime pay might boost consumer purchasing power, spur production, and put more people back to work.[32]

The signing of the FLSA on June 14, 1938, with Wagner again at FDR's side, marked the culmination of six years of progressive legislative measures to address the needs of workers, both those with and without jobs. Despite the attrition of conservative southern Democrats and the often conditional support of organized labor, the now-wobbly New Deal coalition could nonetheless boast a formidable new body of laws that emboldened the worker in his negotiations with management (the NLRA), provided a minimum wage,

and established decent working hours (the FLSA). Moreover, Wagner's earlier inclusion of unemployment compensation in the Social Security Act
helped to finally create a stable, if not always adequate, safety net for those
that lost their jobs. The administration's bold community improvement projects and public works programs, through FERA and the WPA, also provided
jobs for millions of unemployed workers. The BLS was gradually improving its count of the unemployed, and with an enhanced Employee Services
Administration, the states were sharing information regarding job opportunities. Looking back, since Harding's Conference on Unemployment in 1921
and Hoover's call for businesses to enlist their communities in associations of
support for the unemployed, the nation had moved quickly beyond the sort
of ad hoc, voluntary response to unemployment that had long characterized
the nation's industrial history. In just over a decade, the liberal-labor coalition
had indeed erected a fabric of laws that attended to this long-ignored constituency. And yet, with technology constantly making new incursions onto
the factory floor, the efforts somehow seemed inadequate. The liberal-labor
coalition continued to urge the creation of some institutional mechanism
that might accurately anticipate the next technological turn, and by doing so
avoid the next wave of unemployment.[33]

Coming Full Circle

Roosevelt was now among the most ardent in expressing frustration over the
insidious impact of technology as it continued to erode jobs in unpredictable
ways, decimating unsuspecting industries and communities. Ironically, the
decade that began with Hoover's Committee on Social Trends highlighting
the role technology was playing in a flagging economy would now end with
the formation of a Temporary National Economic Committee (TNEC) that
would ask itself the same question. In requesting the committee, Roosevelt
also returned to a familiar theme, warning about the dangers of powerful
monopolies that distorted competition and fixed prices and production.
He thus asked Congress to make a thorough examination of "the effect of
that concentration upon the decline of competition." By July 1938, Congress enacted a resolution to form the special committee, which comprised
a bipartisan balance of congressional members and an equal number of
officials from the executive branch. Wyoming Democratic senator Joseph
O'Mahoney, who was selected as the committee chairman, possessed a reputation for being wedded to the facts. He was generally considered a New
Dealer, though he vigorously opposed FDR's court-packing scheme and was
not afraid to take issue with the administration when necessary. Above all,

he was known for being a tough-minded investigator, dating back to his editorship of the *State Leader* in Cheyenne and later his time as a practicing attorney in Washington, where he helped unmask the preferential treatment of lessees of the Teapot Dome Naval Oil Reserves.[34]

At the very beginning of the TNEC's deliberations, O'Mahoney turned to one of its own members, Isador Lubin, now in his sixth year as the director of the BLS, to summarize exactly where the economy stood at the end of 1938. As the beneficiary of Wagner's legislative push to expand the bureau's role, "Lube," as Wagner affectionately referred to him, had worked with Bernard Baruch at the War Industries Board during World War I before joining Wagner's staff. He was a tireless worker, and once he was appointed head of the BLS in 1933, it took little time for him to catch the eye of FDR. He quickly became a de facto member of Roosevelt's brain trust, navigating between the White House and his director's office. Though Roosevelt often referred to him as his favorite economist, that did not prevent the professional civil servant from painting a dire picture of the economy, even to FDR himself.[35]

Thus, at the very outset of the TNEC hearings, Lubin's graph-filled, day-long survey of the economy clearly illuminated the troubling realities facing policymakers. Americans' per capita income was less than it was in 1932, and in 1932 it was less than it had been in the previous decade. Since 1929, the United States had lost some $133 billion in national income when adjusted to a fixed price level. The employment picture Lubin described was just as bleak. Both the capacity to produce and the efficiency of production increased throughout the Depression, but they did so without commensurate job growth. Almost seven million US households, or 22,300,000 individuals, were part of either a works program, an emergency program, or direct relief. In short, New Deal programs, rather than the private sector, were providing for too large a portion of the working population. Though laborers could take some comfort in the impressive array of new laws in place to protect them, there was little new job growth in the still-ailing private sector.[36]

As its charter required, the TNEC devoted itself at the outset to studying the impacts of monopolistic practices. But by the spring of 1940, it focused on unemployment resulting from technology. The testimony followed familiar and predictable lines, with some arguing for technology as the solution, some seeking to tame it, and some demanding political redress. "Change in one [technology] must be synchronized with change in the other [policy]," urged Theodore Kreps, the committee's lead economist, seeking to offer some hope that there might be a balance struck with the right set of new economic policies. William Green and other union representatives, however, had lost all patience. Green told the TNEC unequivocally that "the crux of the problem we are facing today is the relationship of technological developments to employment opportunities." Across the economy, he claimed, technology was shown to be clearly taking more jobs than it was adding. "The labor content of our basic products has been declining," Green stated. And he was quick to add that companies were not sharing the increased revenues from enhanced productivity with their workers in the form of higher wages. Green's comments ran headlong into the predictable retort of those still clinging to their techno-optimism. Thomas J. Watson Sr. unabashedly extolled the positive aspects of the new office technologies that his company, IBM, pioneered. Watson noted that between 1929 and 1939, the number of workers operating these new machines that counted, computed, typed, and recorded increased by some 300 percent, and the annual wages IBM paid increased some 30 percent. Watson believed that as office operations became more sophisticated, and machines produced more information, the number of employees needed to process and analyze information would likewise grow.[37]

Moving beyond this predictable rhetoric on either side of the technological unemployment debate, the TNEC completed its own in-depth analysis of over a dozen industrial sectors. As with Weintraub's earlier analysis, it was unable to come to some determinative answer on how corrosive technology was to employment. The issue of technology's overall role remained unsettled, and the TNEC tried to express its ambivalence over the unresolved question by genuflecting to both sides of the argument:

> On the one hand it is still claimed by some that the factor which is of special significance in making the problem of unemployment what it is today is that of technology. Not only the number of the unemployed but their distribution by occupations, the duration of their unemployment, and their chances for being reemployed are presumably influenced in large measure, if not primarily by the technological changes which have been taking place during the past two decades and which promise to continue into the discernible future. On the other hand, there are those in this country as well as abroad who either deny the existence of technological unemployment entirely or regard it as of minor importance.

Nonetheless, the TNEC did note a precipitous decline in overall man-hours, from 19.9 million in 1929 to 14.9 million in 1939. The adoption of new processes in steel, the increase in the speed of spinning techniques in textiles, and the substitution of machines for men in the tire industry were among many examples the committee identified where the introduction of new technology diminished any beneficial impacts to overall employment accrued from mandatory limits on work hours. Moreover, the committee noted that displaced workers were facing longer time horizons between jobs, the burdens falling hardest on older workers, and it cited an increasing need for training those whose skills were obsolete.[38]

After years of New Deal price and wage adjustments, and the emergence of the new Keynesian fiscal theory that allowed government more latitude in the use of its own budget to spur demand, it was clear to the authors of the TNEC report that the rate and concentration of capital expenditure on new equipment was increasing the pace of technology entering the workplace. As Senator O'Mahoney concluded the committee's hearings, he again called on Lubin to leave his seat on the dais to provide a final perspective on the role technology played. Lubin began by cautioning that other factors, such as cheaper labor, local tax conditions, and the sheer exhaustion of available resources, also had a role in the erosion of jobs. In some cases, these additional forces, coupled with the introduction of new technologies, were combining to eliminate entire communities. This perfect storm hollowed out towns in the New England textile industry and the upper midwestern timber industry. As Lubin made abundantly clear, it was advancing technology that remained the most pernicious of the causes of continued high rates of unemployment. Its job-destroying impacts offset any advantage the economy gained from having a wider variety of less expensive goods, and more profitable enterprises. In what now seemed an endless Depression, the reality of unemployment, and technology's large role in causing it, appeared to be the economic norm. In the context of growing international tensions, Lubin's frustration echoed a similarly candid assessment offered during the hearings by two Brookings economists. They, too, seemed to have no answer. Frustrated, one of them rhetorically asked, "Where else can the stimulus be expected to come? From war?"[39]

CHAPTER 3

Technology's Triumph

They don't want to come back to chaos. They want to come back to a job.

—Henry M. Rosenberg, plant manager, Electronics Corporation of America, testimony before the Senate Committee on Military Affairs, June 9, 1944

Throughout his first two terms, Roosevelt relentlessly delivered on a series of progressive initiatives aimed at stimulating employment and assisting the unemployed. But as he neared the end of his second term, many experts inside and outside the administration were convinced that the steady advance of machines in the workplace prevented a sustained recovery. Though it remained a challenge to precisely measure technology's role in the unrelenting unemployment crisis, the Temporary National Economic Committee concluded that it was a central actor. Steady improvement in productivity driven by automatic machinery was overwhelming the many efforts aimed at putting people back to work. In his State of the Union address in 1940, Roosevelt euphemistically conceded, "We have not yet found a way to employ the surplus of our labor which the efficiency of our industrial processes has created." Though an advancing war in Europe presented ample opportunity to increase industrial capacity, Roosevelt seemed hesitant to advocate for a massive war buildup. In his annual message to Congress, he rejected what he called "the European solution of using the unemployed to build up excessive armaments, which eventually result in dictatorships and war." Roosevelt, who prior to 1940 steadily bucked the headwinds of isolationism, was nonetheless reluctant to use the call for military preparedness as an excuse for economic stimulus.[1]

At the end of April 1940, Gill and Weintraub, who earlier had authored the WPA's National Research Project, told Congress that the economy now seemed to be stuck in a "vicious cycle." They believed that the lag in demand for capital goods was the greatest impediment to a full economy-wide recovery. Though the costs of production declined, the reduction was not translating into business investment in new equipment. As a result, they urged an expansion of government investment to boost production on a grand scale. They noted how earlier in the decade, the government relied on the funding of public works as the principal means of stimulating consumer spending power as well as industry's purchase of heavy equipment used for construction. Yet any positive effects from investment in works projects quickly dissipated. Nonetheless, Gil and Weintraub maintained that yet one more wave of capital investment was what was required to spur production and employment. Investment in new machinery remained the critical means for jump-starting a sustained recovery. Yet in the same breath, they added a caveat to their claim by acknowledging that it was these advanced technologies now operating the nation's factories and increasing labor productivity that were causing the unemployment crisis. The cure they proposed, it seemed, was the very cause of the problem, and the two experts had no solution for escaping the vicious cycle they described.[2]

Seemingly at the end of the road of new solutions, as Roosevelt headed into his third term, an astounding 50 percent of the nation's manufacturing capacity remained idle. And yet, Roosevelt's constant attention to the unemployment problem, combined with his ebullient personality, continued to resonate with the US electorate. His patrician aplomb could not obscure his heartfelt empathy for the jobless. And his tireless efforts to find new ways to break the cycle of unemployment connected with those millions still without work. Even the Republican presidential candidate, Wendell Willkie, seemed aligned with FDR on most of the important issues. The size of Roosevelt's 1940 victory—he won thirty-eight states and convincingly defeated his opponent by some 10 percent—seems all the more remarkable in an economy continuing to falter, with the unemployment rate hovering at 14 percent. But as he embarked on his third term, Roosevelt found that the very same technology that had proven such an impediment to his New Deal initiatives suddenly would become the enabler of a dramatic and energetic recovery. A nation that for over a decade had experienced so much unemployment was about to enter an era of full employment. Technology, the unrelenting force that so often took jobs, suddenly became the engine of conversion to a wartime economy and the panacea for an ailing economy.

In the dramatic, seemingly overnight redesign of factory floors to meet the exigencies of war, and then in the just-as-efficient reconversion back to peacetime production, a new techno-corporate model emerged that not only enabled the Allied victory but also demonstrated the value of efficient economic planning. In the altered post–New Deal politics, the liberal-labor coalition would seize on that model in its attempt to sustain a postwar full employment economy.[3]

The Palliative of Conversion

During a post-1940 election cruise in the Caribbean, Roosevelt read and reread a letter he had received from Winston Churchill. The besieged prime minister later said that the letter was among the most important he ever wrote. So persuasive was Churchill's case for war aid from the United States that when FDR returned to Washington, he proposed the Lend-Lease plan, a first step toward mobilization. While up to this point the president had resisted the temptation of using a military buildup as a stimulus for recovery, he had carefully laid the groundwork for intervention not only with his words but also by quietly putting in place an administrative apparatus that could quickly ramp up a war-mobilization effort when necessary. Beginning in August 1939, he quietly created a War Resources Board to advise the armed services regarding plans for mobilizing the economic resources of the nation in the event of war. Then in 1940 he signed the Selective Service Act, establishing the draft. Accompanying his January 1941 call for providing Europe with arms, he created an Office of Production Management to facilitate conversion of industrial capacity, a Defense Contract Service to expedite the contract procedures, and a Supply Priorities and Allocations Board to accelerate the procurement of essential raw materials. His Lend-Lease proposal quickly passed in the House on February 8, 1941, despite the strident pleas of isolationists. Roosevelt's political skills in continuing to cajole those even within his own ranks who resisted US intervention, including his ambassador to England, Joseph Kennedy, led to final Senate passage on March 11. Rather than being stimulated with production targets, wage controls, or public works projects, the economy now responded to the stimulus of war preparation. The fight, as Roosevelt defined it, was for the very survival of democracy, or, as he couched it, "the Four Freedoms," including the right to a decent job.[4]

The steep increase in military spending that began in mid-1941 finally turned the unemployment needle in the opposite direction from where it had seemed stuck since 1929. After the attack on Pearl Harbor on

December 7, 1941, the specter of continued high rates of unemployment quickly faded. In his January 1942 State of the Union address, Roosevelt called on Americans to seize the technology advantage. "The superiority must be overwhelming," he urged. "We have the ability and capacity to produce arms not only for our own forces, but also for the armies, navies, and air forces fighting on our side." Setting forth specific targets for planes, submarines, tanks, and so forth, the mobilization took on such momentum that by March 1942, unemployment had been cut in half from March the year before, declining from 6.6 million to 3.5 million. Almost overnight the auto industry converted from making autos to producing armaments. Throughout the country, factories previously undermanned, due to the simple lack of demand for consumer goods in a depressed economy, now found themselves frantically searching for enough employees to fully supply the escalating demands of a war economy. Not only was the mobilization demanding more skilled and unskilled laborers, but the armed services were taking workers out of the factories and demanding that new recruits quit their jobs immediately, thus leaving others, like the older unemployed and women who had never considered working a factory job, to fill their shoes. Concerns about unemployment suddenly seemed a distant memory. One columnist remarked, "The war has finally accomplished most of what the New Deal set out to do." Technology, so disparaged in the previous decade, now was the force helping democracies to prevail against fascism.[5]

Initially businesses seemed reluctant partners in the president's call to assist the Allies through Lend-Lease. In contrast stood the unbridled enthusiasm of the young vice president of the United Auto Workers, Walter Reuther. Reuther grew up in Wheeling, West Virginia, as one of five children in a German immigrant family. In 1927, at the age of nineteen, he set out for Detroit, where, as indicative of his persuasive skills. he remarkably convinced his supervisors at Ford that he qualified as a tool-and-die maker even though he lacked the substantial schooling and training typically required. Reuther was enterprising, resourceful, and convincing. Not only did his capacity for persuasive argument earn him this coveted position, but it also quickly elevated him within the ranks of the United Auto Workers. A year before Pearl Harbor, as a rising star in the labor movement, he offered his own Reuther plan, intended to utilize the auto industry's excess capacity for the production of military aircraft. Reuther's bold vision was a harbinger of what would soon be the wholesale conversion of major sectors of the atrophied US industrial machine for wartime production. His plan called attention to the staggering deficit in machine tools production that plagued the auto industry and many others.[6]

With the Japanese attack, Reuther's plan to scale up industry in preparation suddenly looked prescient. By mid-1942, with the nation on full war footing, the formerly reluctant private sector patriotically pivoted to embrace its new partnership with the federal government in the full-scale production of arms and the development of advanced new weapons of war. Accompanying the emerging alliance between the public and private sectors came sophisticated managerial and governmental bureaucracies with wartime disciplines that would soon permeate the social fabric. New machines of war—from tanks to B-1 bombers, from machine guns to destroyers, and from landing craft to field radios—were being produced at a rapid pace in now fully converted and fully manned factories. Military production that ramped up gradually under Lend-Lease increased exponentially after Pearl Harbor. Analysis of the shifts in manpower reveals that five sectors of the economy alone (agriculture, retail trade, wholesale trade, construction, and motor vehicle production) needed the equivalent of over one million full time employees (FTEs). For example, the transportation equipment industry alone, during the two- year period that ended in 1943, acquired some 2.6 million FTEs, a 384 percent increase over 1941 employment levels. By the end of 1943, industrial production of durable and nondurable goods was double 1939 levels.[7]

In the suddenly transformed wartime economy, unemployment was virtually eliminated. The mobilization for World War II would require every American, including those previously excluded—namely, women and African Americans—to become, at least temporarily, part of a rejuvenated industrial sector. Not only did unemployment vanish, but the demands of new wartime production caused the total workforce to increase by some ten million, over half of whom were women. So encompassing was the war mobilization juggernaut that by the end of 1944, it was estimated that only about 250,000 men between the ages of twenty and sixty-four remained unemployed. At the same time, the number of women in the workforce increased by about one-third between 1940 and 1944 just to keep pace with the demands.[8]

With what seemed lightning speed, the Depression era's fixation with technological unemployment gave way to a new optimism about technology's critical role in securing an Allied victory. While soldiers in the field revealed heroism and bravery that won deserved plaudits, the role of technology in winning the war received its own share of accolades. On October 13, 1942, in a hearing before the Senate Military Affairs Committee, Senator Harvey Kilgore (D-WV) told his colleagues, "This is a technological war. It is a war of machines and men. It is a war of science. To win such a war, it is obvious that we must not only mobilize men, but we must also mobilize technology." John Norris of the *Washington Post* wrote, "Courage and fighting ability still are important, but the side with the best weapons decidedly has the edge." As he noted, the sum total of new technologies "will turn the tide in this machine-age war." Technology, the bogeyman of the last decade, was now being cast as the savior of democracy. Away from the front lines of war, as industrial production hummed along at full tilt, stood an impressive cast of scientists who were recognized for their critical role. The war not only bolstered the reputation of technology writ large, but also enhanced the reputation of scientists whose discoveries now translated into victory.[9]

The Patriotic Scientist

By World War II, the scientists Norbert Wiener and Vannevar Bush had gained substantial reputations among their peers. Though both would accept teaching jobs at MIT in 1919, their paths seldom crossed. Both had also been undergraduates at Tufts, but the child prodigy Wiener graduated thirteen years before the older Bush arrived. Their careers then moved in entirely different directions. Wiener was the pure scientist. Just as the Scottish botanist Robert Brown had pondered the dilemma of the seemingly random motion of airborne pollen over a century earlier, Wiener applied Einstein's theory of molecular motion to the haphazard motion of electrons through telephone wires and from radio towers. His paper on the subject, entitled "Generalized Harmonic Analysis," appeared in an obscure Swedish mathematics journal and went largely unnoticed. But his insight provided electrical engineers a way to explain and thus control the seeming randomness of irregular currents of all sorts and, usefully, to explain the chaos interfering with radio transmissions. As one contemporary described it, he seemed in one fell swoop to have explained all of the wave problems that had long bedeviled communications engineers. Perhaps more significant, Wiener's mathematical insights also led him to develop a new statistical approach to communications engineering that would soon transform the workplace by providing a platform for the new technology of computers.[10]

Throughout his scientific career, Wiener largely ignored the commercial potential unleashed by his discoveries. In contrast, Bush, in the American tradition of pragmatic inventors from Franklin to Edison, was enamored with the practical side of scientific inquiry and its commercial possibilities. Thus, while Wiener was examining the randomness of radio waves, Bush was attracted to possible real-world applications of the new radio science. In his early days as a professor, he quickly became involved with the American Radio and Research Corporation (AMRAD). The start-up enterprise housed on the Tufts campus was backed by J. P. Morgan. By 1915 it began occasional broadcasts months before the nation's first station, KDKA in Pittsburgh. Unlike Wiener, Bush may not have been considered a scientific or mathematical genius, but his prodigious mind was always devising some new idea that he could lateral on to other researchers for further development. He also had a keen eye for early discoveries that might have practical use. His AMRAD experience taught him valuable lessons about the complicated odyssey that new discoveries encounter on their way to commercial application. Similarly, during World War I, though he unsuccessfully attempted to sell the US Navy on a deep-sea detection device, the experience left him with a better appreciation of how private entities might partner with government in research and thus share in the commercial development of new discoveries. Though as early as 1922 AMRAD was going bankrupt, the firm's patents, largely Bush's progeny, were purchased by a new company called Raytheon. The new radio tube technology Bush developed quickly reduced the price of radios. As the radio business joined the parade of other industries booming in the twenties, Raytheon sales quickly topped $1 million.[11]

Neither Bush's initial commercial exploits nor Wiener's breakthrough discoveries brought either much public renown before World War II. Wiener's masterwork, *Cybernetics*, detailing his mathematical language that helped machines talk to one another, would not be published until 1948. As for Bush, though the fledgling Raytheon became a dependable supplier of superior radio S tubes, initially it had a limited public profile. Meanwhile, Bush was perhaps more recognized in scientific circles for developing a cumbersome analog computing device he called the differential analyzer. It was meant to solve differential equations by varying electrical currents while the operator physically interchanged the machine's mechanical components. Before the war, Wiener's original research and the evidence of Bush's developmental genius were apparent only to a select number of mathematicians and scientists. Wiener would remain a largely unrecognizable figure publicly, even though he would later be associated with birthing the new automated postwar economy. Bush, on the other hand, would become widely recognized as the scientist who mobilized technology to win the war.[12]

Bush's rapid ascent in leading the United States' scientific army can be traced to the exciting new initiatives he took in the emerging technology of microfilm, which propelled the research for what would eventually become sophisticated encryption technology. His so-called rapid selector devices, which he continuously improved in the late 1930s, caught the eye of the US Navy. His growing reputation as a practical scientist emerged not only from his creative research initiatives but also from his adroit networking with scientists and engineers at the forefront of emerging new aspects of information and computer technology. By 1939, Bush's reputation as a leading scientist who was also an effective manager with a wide range of contacts across the scientific community led the Carnegie Institution of Washington, among the nation's leading scientific organizations, to appoint him as its president. Bush's already impressive network of leading-edge scientists and technologists quickly widened.[13]

In his new role, Bush began to speak out on the need for greater war preparedness. He became increasingly convinced that the government lacked a concerted and coordinated approach to supporting and accelerating technological development. In Bush's view, the government did not enlist the private sector sufficiently, and the siloed departments of government further hindered preparedness efforts. In June 1940, after the German invasion of France, and following a series of consultations with those in his burgeoning network, Bush presented Roosevelt with a plan to coordinate and accelerate the nation's technological war effort as part of an overall preparedness strategy. Impressed with Bush's blueprint for mobilization, Roosevelt's chief of staff, Harry Hopkins, condensed the plan into a single-page memorandum. Aware of Bush's reputation, Roosevelt hardly needed much convincing. The president promptly named Bush to head the Office of Scientific Research and Development (OSRD) and the two quickly struck up a very close relationship. Bush's impact was immediate and obvious. He drew on his business acumen and his relationship with the leading companies of the era to dramatically accelerate the pace of technological development. He was able to convince his vast network of fellow scientists to put down their own projects and pitch in on the war effort. As head of the OSRD, Bush skillfully managed what the *New York Times* described as "more than a hundred thousand trained brains working as one." While certainly the *Times* employed journalistic wartime hyperbole, the mobilization of science in such a brief period created an impressive range of new weaponry. In addition to new developments in conventional weapons technologies, there were also new medical advances, including antimalarial drugs, blood substitutes, and a capability to produce penicillin on a safe and effective scale.[14]

Bush's OSRD skillfully sought out allies in the Department of War and convinced them that even the young scientists eligible to enlist were more valuable to the war staying stateside. Those on the front lines who understood how important the development of new weaponry was to victory quickly reinforced Bush's plea to Pentagon leadership. His multifaceted and intensive efforts to stymie the technological advantage of the Axis powers soon resonated not only through the halls of the Pentagon, but also with allies. When the Germans launched their new robotized missiles on England in the spring and early summer of 1944, the V-8s were largely foiled by a combination of a sophisticated new form of radar, new electrical sensing devices, and radio-activated explosive devices, all invented and made in the United States at the behest of Bush's OSRD. General Sir F. A. Pile, chief of the British Antiaircraft Command, sent Bush a copy of the report he wrote about the largely successful British counter to the German attacks, with the handwritten note, "With my compliments to OSRD who made the victory possible."[15]

Among the many individual soldier scientists Bush enlisted in the cause, Wiener himself devised an effective antiaircraft capability that could anticipate the wobbly flight of wounded or disabled incoming aircraft. Though Bush actually placed more confidence in a rival corporate effort, Wiener's far less expensive design proved to have far better results. Of course, true to his reputation as the pure scientist, Wiener seemed prouder of the publication of the 120-page paper that accompanied his radar device. In it he detailed his approach to integrating the emerging field of communications science with physics. The paper would form the basis of his *Cybernetics*. With it, he redefined how scientists and engineers would see converging worlds of machines and signals. Wiener mathematically prescribed how radio waves, telephone signals, and radar beams could effectively control machinery, whether for war or industrial production. As one of Wiener's biographers later noted, "In a surgical cut, he severed the entire practice of control engineering, which had been the province of power engineers historically, and brought it bodily into the camp of communication." Thus, in making his singularly large contribution to radar defense, Wiener was laying the groundwork for what would become an entire information economy.[16]

Except for its important wartime application, however, the true potential of Wiener's discovery remained invisible and indecipherable to politicians. By contrast, Bush's efforts produced tangible results for all to witness and acclaim. The most dramatic OSRD initiative, the Manhattan Project, ended the hostilities in the Pacific and further bolstered Bush's growing public reputation. Yet because of his focused management of the nation's scientists, Bush exhibited little tolerance for any distracting narratives. When he was presented with an honorary doctorate at his alma mater, he used the occasion to question labor officials who led strikes at the nation's weapons plants. Having watched the democratic nations of Europe fall one by one to the Nazi military juggernaut, Bush was incredulous that "we should see in this nation shipyards and airplane factories shut down because a few men cannot agree on things that seem by comparison inconsequential." In Bush's view, these were "petty bickers" that unfortunately distracted from the daunting task of converting the economy to full war footing. If scientists were being enlisted in the war's cause, Bush argued, then certainly the unions should do their part.[17]

Emerging Fault Lines

Bush's attitudes toward labor suggested larger tensions that surrounded the labor movement during the war. While management and government partnered over ramped-up production schedules, organized labor often felt like the odd man out. Since the unions stood at the core of the liberal-labor coalition, maintaining their public credibility was critical to the political coherence and strength of continued initiatives on behalf of laborers. Yet during the war, organized labor was not always in lockstep with the administration's goals. Moreover, adding to the tension, Roosevelt had consistently been more the advocate for the individual laborer than a champion of their unions. On the one hand, Roosevelt recognized and politically benefited from the growing political clout of organized labor and sought to maintain the allegiance of its leadership. On the other hand, his messages and actions were typically aimed at the rank-and-file laborer. Union leadership was not oblivious to Roosevelt's schizophrenic position. Even prior to the war, union agitation over the details in legislation such as the NLRA or FLSA suggested that union leadership's real concern was about being taken for granted.

The sheer numbers worked in Roosevelt's favor. As the conversion to war production moved ahead with breathtaking speed in US factories, at least two of every three workers were not card-carrying union members. Moreover, union political influence was limited primarily to states above the Mason-Dixon line. Yet these realities did not deter northern union workers from arguing that their wages did not keep up with inflation. Attempting to dutifully exercise their patriotic support for the war effort by making sacrifices to work longer hours for less money, AFL and CIO members were asked by the government to take "no-strike pledges," a sacrifice that soon tested the patience of union leadership. In 1943 a spate of strikes broke out at defense plants around the country, sparking an immediate public backlash and expressions of outrage in Congress. Concerned that such continued work stoppages could hamper the war mobilization effort, FDR established a National War Labor Board (NWLB) to mediate these disputes in hopes of containing the spreading tension, and Congress quickly enacted the Smith-Connally Act (1943), which gave the president greater wartime power over unions and the ability to curtail strikes.[18]

While Roosevelt was busy tamping down the rash of strikes, labor leaders felt no trepidation about continuing to argue with management over hours, wages, and conditions, even as their membership attempted to fulfill their patriotic duty. In an effort to quell the unrest and growing public resentment, the AFL's William Green took the lead in forging a wartime compromise on labor wages, known as the Little Steel formula. Roosevelt's NWLB quickly adopted the terms of the agreement as the administration's template for wage controls throughout industry. The formula allowed for some upward movement in wages in order to keep pace with inflation. Nonetheless, Green's efforts to retain some modicum of wage growth were not viewed as satisfactory by all labor. Within labor's ranks, the suspicion grew that the obviously strong alliance between business and government, forged to facilitate mobilization, was giving business the upper hand in exacting favorable labor contract terms. During the Depression, industry had complained about the NRA's heavy-handed price and wage controls and felt that labor had the advantage. Now it was labor's turn to voice its concerns over the NWLB's power over wages, and no labor leader was more forceful in expressing displeasure than CIO president John L. Lewis.[19]

Lewis, whose newly formed CIO was so helpful to Roosevelt's 1936 reelection, was bitterly opposed to Roosevelt's interventionism. Taking a strongly isolationist position prior to the war, he aggressively encouraged his members to demand greater wage increases and then rejected the Little Steel formula in the spring of 1943, sending some half a million workers out on strike. His belligerent tactics so irritated FDR that the president called for the temporary government takeover of the nation's coal mines. While Lewis agreed to a cooling-off period, wildcat strikes continued. But the CIO strikes fueled popular animus against labor. Though Roosevelt was presented with an opportunity to quash labor dissent, politically he understood the importance of the unions' continued role within the liberal-labor coalition. Labor also retained enough political influence over the president that it had succeeded in convincing him to veto the Smith-Connally legislation effectively limiting strike actions. In exercising his veto, however, Roosevelt realized he could enjoy the best of both worlds. With his veto pen he maintained labor allegiance, knowing all the while that sentiment ran so high against Lewis's belligerence that Congress would almost certainly override the veto. Indeed, that is precisely what Congress did. Yet during the war years, in spite of the antistrike law, new labor walkouts would continue to test FDR's nerve, as well as the patience of other notable public figures. The otherwise quiet and restrained General George Marshall went so far as to suggest that labor's stubbornness added another six months to the war.[20]

Adding to the tensions that wartime labor strikes aroused was the increasing political anxiety over the anticipated influx of GIs who eventually would return to the workforce. Millions of GIs would soon return home and compete for jobs in a more efficient and productive industrial economy. After one dramatic confrontation at an aeronautics plant in May 1944, Donald Nelson, who chaired the War Production Board, tried to sympathize with the domestic workforce. He warned that "what gives workers and others 'cutback jitters' is the fear that they will find themselves thrust into a planless and chaotic economic state of affairs as soon as war orders fall off on a large scale, becoming worse when peace comes." An exhausted Roosevelt soon began to fear that without a common enemy, Americans facing the unknowns of reconversion might turn on one another. He was particularly concerned that in a transitioning economy, labor would turn on its own, or worse, that the temptation for racial strife would boil over. Just as in his Four Freedoms speech in January 1941, FDR served notice of how freedom from the fear of want should be honored in an expanded democratic state. In putting forward what would be called a Second Bill of Rights, Roosevelt was attempting to reassure US workers that in a postwar economy there would continue to be enough jobs for all. With its pledge to ensure "the right to a useful and remunerative job," Roosevelt hoped this Second Bill of Rights might alleviate any concern that a postwar economy would revert to prewar conditions. Indeed, Roosevelt envisioned the model of full employment established by wartime public-private cooperation as continuing.[21]

The challenge of sustaining full employment in the postwar economy involved not only the influx of millions of returning GIs but also new demographic realities. As the war progressed, African Americans streamed to the industrial North from the rural South, attracted by newly available defense industry jobs. In the wartime economy, the black population in California quadrupled. Leaders such as A. Philip Randolph, well understanding the long history of black exclusion, pressured FDR to ensure that African Americans received their fair share of domestic defense industry jobs. When Randolph threatened a massive march on Washington, FDR took action to remedy the situation. Similar demands continued throughout the war era, exacerbating white workers' anxieties, and in some instances leading to violence. In June 1943, racial tensions mounting for weeks in cities like Detroit and Los Angeles resulted in uprisings that left thirty-four dead and one thousand wounded. Foreseeing the potential for continuing racial violence to manifest itself in the postwar world, FDR used the occasion of National Brotherhood Week in February 1944 to call for maintaining a sense of shared unity. "On land and sea and in the air, the sons of the United States fight as one though they come from every racial and cultural strain. And though they worship at different altars, they are brothers in arms now. Soon, pray God, they shall be brothers in peace." Though the president sought to raise awareness of the considerable sacrifice and heroism of African American soldiers on the front lines, commensurate actions that might have accelerated an integrated postwar society were less forthcoming.[22]

Simmering racial tension was not the only challenge facing the liberal-labor coalition in the postwar era, threatening the continued cohesion of this formidable alliance so disciplined in its support of New Deal policies. The new politics of reconversion was already forcing organized labor, with its tattered wartime public image, to recalibrate how best to wield influence. In the new political climate, technology no longer would serve as the default bogeyman it was during the Depression. Scientists and engineers now inherited the political windfall from technology's wartime contributions, and their expertise assumed a more influential role in the halls of government. Bush himself seemed the embodiment of a new inventive and innovative elite that freely traversed the boundaries between government and the private sector. Additionally, though they often sparred over how to unwind the contractual terms of war contracts, corporations and government seemed as closely aligned as they had been in their disciplined effort to win the war. A new corporate emphasis on research and development fueled by wartime technological prowess was a tangible result of the new postwar business-government alliance.[23]

Faced with a strikingly different postwar political landscape, labor leaders thus returned to a familiar narrative, emphasizing the need for business to share the benefits flowing from the plethora of emerging technologies and improved productivity, thus lifting the living standards for the returning blue-collar laborer. Union leaders had little option but to join the public chorus in praise of the United States' technological juggernaut. In doing so, union leadership discovered that it not only had allies for its view within the administration, but had an entire organization dedicated to sustaining full employment after the war ended. As early as January 1943, Alvin H. Hansen, a Harvard economist working for the National Resources Planning Board (NRPB), was already exuding an optimism about the role technology might play in a postwar world. The NRPB, created by an executive order in 1934, was a holdover from the New Deal and a few months away from extinction. Yet for some time it had moved away from its original mission of choosing new public works project sites and toward the consideration of broader questions of how best to construct a postwar economy. The NRPB issued a series of reports, spiked with heavy doses of the new Keynesian economics, that stressed how tax and fiscal policy, as well as the construction of a more robust social safety net, might become the new normal in a highly productive postwar economy and thus maintain the full employment enjoyed throughout the war. In its report *After the War—Full Employment* (1942), NRPB economists provided an optimistic assessment: "We shall have when the war is over, the technical equipment, the trained and efficient labor, and the natural resources required to produce a substantially higher real income for civilian needs than ever before." Thus at the very outset of the United States' conversion to a wartime economy, the NRPB envisioned sustained full employment after the war. The board's analysis provided the liberal-labor coalition with substantive ammunition as it pursued legislation that would ensure that such predictions became reality.[24]

Reconversion Politics

In the transitional postwar period, however, a fear grew among liberal-labor reformers that in the postwar economy, large corporations would pursue higher profitability by dramatically reducing from their wartime capacity. The specter of vast layoffs grew, and so too did fears of a return to Depression-like conditions. Given the terms of wartime contracts, labor was anxious that some companies might shut down entirely for a quarter or more while they retooled and reconfigured their operations. Indeed, many in business were busily unwinding their agreements with Uncle Sam and offloading wartime surpluses. The federal government technically owned almost two-thirds of the nation's machine tools, as well as manufacturing facilities estimated to be worth some $9 billion that during the war produced essential basic commodities such as rubber, steel, aluminum, chemicals, and aircraft. During the frenzied mobilization effort, the entire face of US manufacturing had been totally transformed from its prewar identity. Plants once devoted to making playground equipment turned out gun mounts. An appliance maker produced lifeboats. A textile plant made grenade belts. As the war wound down, the reality set in that the United States' entire industrial apparatus would have to return to productive peacetime footing and restaff plants. Many of the wartime factories teemed with men and women producing war supplies, but others operated quite efficiently with only a few laborers. Notably, the successful New Deal creation, the Tennessee Valley Authority's Oak Ridge facilities, which had been converted to produce chemicals, petroleum, rubber, and atomic materials for wartime application, operated with miles of instrument panels controlled by only two hundred operators and virtually no unskilled laborers. Fears grew, particularly among organized labor, that similar machine-induced efficiencies would become the new standard in the postwar economy.[25]

"Conversion without Depression"—namely, how to avoid it—was the title of a panel hosted by the *New York Times* five months before the war ended. Moderated by Arthur Krock of the newspaper's Washington bureau, labor leader Reuther began by provocatively resurrecting the idea of a thirty-hour workweek as one way to help sustain full employment once the war ended. His opening salvo immediately engendered criticism from Krock's two other panelists, Eric A. Johnston, president of the US Chamber of Commerce, and Henry J. Kaiser, a ship builder who was on the verge of expanding into steel, aluminum, autos, and health care. While Kaiser believed that management should take the lead in directing reconversion, Johnson argued for a larger role for government. As Kaiser dramatically warned, "The shadow of unemployment darkens the rim of the sun." Having turned out over 1,400 ships for the war effort, Kaiser had full confidence that the returning GIs would be fully capable of addressing the challenges of reconversion without the heavy hand of government interference. On the other hand, speaking for the broader constituency of US businesses small and large, Johnson saw government playing an important role in defining the rules for reconversion, notably how to unwind contracts. Perhaps fearing the worst, he also called for the strengthening of state unemployment compensation funds. Reuther was more hopeful that government might command a more direct role to avoid massive unemployment. He concluded that "without Government initiative the effective mobilization of our national economic life and productive effort cannot be achieved." Though each of the panelists argued for full employment as a policy goal, each had effectively staked out the broad contours of the forthcoming debate: how central a role the federal government should play, and what specifically should be tried to maintain full employment in a postwar economy.[26]

All parties to the ensuing debate over full employment agreed that establishing the right level of government engagement was critical. They differed, of course, on how much government engagement was enough. A baseline issue that would immediately impact postwar employment levels was the question of how government should manage wartime surpluses. This was not an altogether unfamiliar challenge. Overstocked inventories exacerbated by a similar overproduction of war supplies from the First World War were thought to be the major cause of the deep recession of 1920–21. Editors of *Fortune* thought the war inventory was so large that for many commodities, the stockpiles could substitute for new production for months, maybe years. Indeed, there were already dramatic examples of the monumental challenges confronting the reconversion effort. In August 1944, *Newsweek* ran a story featuring a picture of a row of rail flatcars mounted with surplus jeeps, with the caption, "A mile of jeeps: Mounted on flatcars, they make a symmetrical picture of the postwar disposal problem." The *Saturday Evening Post* cited examples such as the 187,000 excess soup plates the army now needed to dispose of. Such surpluses would lead to "the biggest bargain sale on earth" unless the government intervened.[27]

For his part, Roosevelt set his sights higher. The surplus issue could be effectively managed; what was imperative was not returning to prewar employment conditions. On this critical challenge he displayed his usual optimism about the prospects for the postwar United States. He pegged the number of jobs that would need to be filled at 60 million, a figure that was comparable to the 54.6 million at work in 1944. Yet Roosevelt declared, "If anyone feels that my faith in our ability to provide 60 million peacetime jobs is fantastic, let him remember that some people said the same thing about my demand in 1940 for 50,000 airplanes." The nation had turned on a dime to mobilize. Now Roosevelt believed it could pirouette again to reconvert. Indeed, the combined efforts that had mobilized the country for war demonstrated how, when technology's full force was unleashed, it could fuel a "full employment" economy. Thus, even before the war wound down, FDR challenged the New Deal coalition, with labor still at its wobbly center, to recast technology's role from villain to savior. Increased productivity no longer needed to be synonymous with unemployment. A new paradigm made productivity an ally in allowing for lower prices, increased wages, and jobs for all.[28]

As the reconversion debate began to take shape, the president was inclined toward what would become billed as the conservative approach. Eschewing his normal progressive instincts, he pragmatically opted for achieving greater efficiencies to ensure a stable postwar economy. Capably disposing of the significant wartime surpluses without creating economic distortions, and thus avoiding thrusting the nation backward into a prewar depressed economy, overrode any ideological preferences FDR may have harbored. The president's corporatist position, however, created an unsettled dynamic within the White House. FDR's own head of the War Production Board, Donald Nelson, adhered to what became labeled the "liberal" approach, since it placed value on the role that small businesses might play rather than relying on large corporations and the now increasingly hegemonic Defense Department. Nelson's approach was an extension of an emerging liberal-labor vision of an expansionist, mass-production, mass-consumption economy fueled by Keynesian fiscal policies. While FDR shared that vision, he disagreed with how Nelson, and for that matter many of the administration's liberal-labor political allies, wanted to achieve it. Annoyed by the dissension within, FDR reassigned Nelson and replaced him with James Byrne, a Supreme Court justice. Byrne was a longtime friend of the president's and a dependable New Deal southern Democrat, an increasing rarity. The transition cleared the way for the administration to adopt the reconversion blueprint laid out by two veterans of the earlier failed World War I experience, Bernard Baruch and his trusted accomplice at the old War Industries Board, John M. Hancock.[29]

Baruch and Hancock's report for the Office of War Mobilization, issued on February 15, 1944, not only was a comprehensive transition guide from a wartime to a peacetime economy, but also carried Baruch's imprimatur, which helped calm the growing anxiety among many in business and government. The two authors took FDR's preferred conservative approach to reconversion, which relied for its execution on the large wartime contractors and sought to achieve economic stabilization in both the short and long term by bringing supply and demand quickly into balance. The reality of severe imbalances stared the country in the face. Beyond dramatically altering what factories produced and creating surplus supplies, the defense buildup necessitated almost seven and a half million more workers than were employed in 1937. Nonetheless, Baruch and Hancock's report sought to reassure the public that it would be easier to reconvert than it was to convert, since the reconversion effort would be eased by other postwar realities. The report predicted that "many women would leave the work force, many returning GIs would go back to school, there would be new domestic and global demands for new products, and many older workers would now retire." An economy straitjacketed into wartime regimens was now free to exert its creativity. Underlying the report's optimism was the belief that the new technologies of war, converted to peacetime purposes, would result in "better houses, better clothes, better food, better safeguards for children, better health care, and wider educational opportunities." With its optimistic undercurrents, Baruch and Hancock predicted that the postwar economy would thrive and prosper. At the center of their confidence lay an abiding faith in the same technological ingenuity that was winning the war. But although the final Baruch-Hancock report received obligatory plaudits from liberal stalwarts in the administration, such as Labor Secretary Frances Perkins, for the progressive columnist I. F. Stone it amounted to what he described as a right-wing economic coup. The fault lines of the debate over full employment legislation thus clearly emerged even before the war ended.[30]

But regardless of whether one was on the liberal or conservative side in the reconversion debate, more fundamentally the widely acknowledged need for a well-coordinated effort between the public and private sectors represented a triumph for the essentiality of a planned economy and the important role that government needed to play in its future direction. Roosevelt's core liberal instincts, as evidenced throughout the New Deal, were to engage government to actively promote the employment and welfare of those the market system failed. His generous notion of the social contract, as defined in his Second Bill of Rights, rendered it imperative that government ensure full employment. If those in the liberal-labor coalition worried that reconversion would only further the increasingly cozy relationship between government and its contractors, it also allowed them to envision how the government might take a central role in planning a future full employment economy. For Roosevelt, the reconversion controversy was not about whether the government should play a lead role in organizing the economy going forward; rather, it was about how that role should be defined. His short-term calculus was to dispense with the transitional issue of surplus property as quickly as possible even if it meant leaving corporations with a central role in expediting the process.

Liberal-labor reformers such as Senator James Murray (D-MT), chairman of the Senate Finance Committee and champion of the call for postwar full employment, disagreed. They did not want the emboldened alliance between corporations and the Defense Department to continue exercising undue influence over the market once the war ended. Murray agreed with Roosevelt that a robust and growing postwar economy necessitated that government continue to exercise its command-and-control authority during the transition. But Murray and his allies wanted to use the first debate over the structure of a postwar economy to serve notice that the power balance needed to change. Their approach to reconversion placed small businesses and individual producers, rather than large corporations, at the center of the business-government alliance. This plan included the direct sale of wartime surpluses to jump-start small businesses and small farms and aid returning veterans. Selling these war goods at a minimal price would raise some needed government revenue, sustain the economy's momentum, and rebalance the economy away from the very "monopolists" FDR had so long despised.[31]

Reconversion Debate

In the upside-down politics of the pivotal postwar debate, Murray and his allies ironically found themselves joined by conservatives such as Robert Taft (R-OH) who supported the liberal approach to reconversion. Taft and his conservative colleagues agreed that preventing large companies from exerting undue influence in the market was the paramount priority during reconversion. They also sought to reduce the need for a cumbersome bureaucracy to oversee the effort. The incongruent political alliance was on display when, on June 16, 1944, the House's and Senate's Post-war Economic Policy and Planning Special Committees that were established to address the issue invited members from other committees that were already deliberating on the topic to attend a joint hearing featuring the newly appointed surplus war property administrator, William L. Clayton. Clayton was Baruch's recommended choice for the post, and he wanted legislation that clarified the rules for dismantling wartime contracts. The government needed to dissolve over one hundred thousand separate contracts with 18,539 firms, which in turn would impact an estimated one million additional contracts with various subcontractors. Clayton argued that the administration needed authority broad enough to allow the government flexibility to address a wide-ranging assortment of different surplus issues. He noted that the government needed, for example, to "sell sleeping bags in Alaska, unwieldy numbers of surplus transport aircraft, machine tools in a Navy machine shop at a Pacific advance base, work in process in a textile mill in Mississippi, and 10,000 pounds of foreign wool." In keeping with the administration's conservative approach, Clayton stated that because of this very diversity, it was highly advantageous that the government's contracts were concentrated among one hundred corporations. Witnesses representing the liberal view disagreed and characterized the administration position as nothing but corporate favoritism. Throughout the hearing, members of the liberal-labor coalition and its conservative allies expressed concerns about giving too much flexibility to officials charged with unwinding these contracts, while Clayton countered that speed and efficiency were the overarching priorities in reconverting the economy.[32]

The heated debate revealed the fragility of the New Deal political alignment in the postwar transition. In particular, the always tenuous relationship with southern Democrats now seemed threatened as a competitive southern industrial base began to emerge. This reality was on full display as reconversion legislation came before the Senate Finance Committee. Like so many committees in both the House and Senate, this critical Senate committee was chaired by a southern Democrat, Walter George (D-GA), who wished to assert greater legislative influence as the war ended. George was no newcomer to opposing Roosevelt when he felt it necessary. During the Depression he consistently opposed federalizing efforts to assist the unemployed, whether through the guarantee of wages and hours or the creation of Social Security. Though outwardly FDR and George seemed to have a civil relationship, the Georgia Democrat often reminded his friends how FDR's political operatives had planted a challenger against him in the 1938 Democratic primary. George predicted that when the president came through Georgia on a campaign swing that year, Roosevelt would "probably take pleasure in kicking me downstairs." Indeed, he was right. When the president stopped in Barnesville, Georgia, on his way to his therapeutic Warm Springs retreat, and with George sitting on the dais, FDR called on the approving crowd to defeat George in the upcoming off-year election. Then FDR went on to praise his own handpicked candidate as George sat helplessly witnessing the entire spectacle. Yet the president miscalculated George's popularity and the senator would go on to defeat his Roosevelt-supported rival. In the ensuing years, George would find various ways to remind the president of his miscalculation.[33]

One such opportunity occurred when George's special committee reported its findings on reconversion issues in 1944. It predictably assumed what was known as the liberal reconversion position, which allowed surplus war materials and plants to be disposed of in a way that was of benefit to smaller businesses and individual consumers. The reconversion liberals were particularly adamant that the plants that the War Department had built with government money and leased to companies to operate should not be turned over to these same companies after the war unless the government received adequate compensation. George's Finance Committee's report urged that the demobilization of plants and supplies be done in such a way as to promote small business, not strengthen monopolies. If excess surplus could not be sold in the United States in a way that was not disruptive, then it should be exported. In the jumbled reconversion politics and to the chagrin of the administration, the conservative George took the liberal reconversion position that "where practicable surplus goods should be sold in quantities that will permit their acquisition by small purchasers."[34]

Further confounding postwar reconversion politics, conservatives in Congress were determined to mount resistance to the same old New Deal approaches to maintaining high levels of employment after the war. Throughout World War II, the Federal Works Administration (the descendant of the WPA), headed by Major General Philip Fleming, continued the complicated public works legacy of the New Deal. During the war and in true Keynesian fashion, the federal government continued its heavy investment in the publicly financed construction of the nation's infrastructure in the name of the overall war effort. At the outset of the postwar economy, this construction continued to provide temporary employment, and in some cases more permanent employment to operate and maintain these new projects. Fleming and his allies maintained that targeted spending on public works in the post-war period helped stabilize the construction industry during this transitional period, with positive long-term employment consequences.[35]

While Fleming thus sought to continue and build on the legacy of New Deal public works spending, conservatives successfully resisted the efforts of an impressive alliance of city mayors and unions that supported funding additional public works projects. Former New York City mayor Fiorello La Guardia, now a progressive Republican congressman, was at the center of those promoting a revitalized public works campaign as the war concluded. But in the 78th Congress, an increasingly more assertive conservative bloc refused to reengage in any sort of massive works effort, particularly one that resembled that of the earlier WPA. Again, as in the case of the Senate Finance Committee, a southern Democratic chairman expressed severe reservations. Representative William M. Colmer (D-MS), like many other powerful southern Democrats, tried to stay with Roosevelt on New Deal policies as often as he could. Yet as chair of the House Special Committee on Postwar Economic Policy and Planning, he was less than enthusiastic about any new round of federally funded public works. When, at La Guardia's request, Bronx borough president James Lyons appeared before Colmer's committee to enthusiastically promote the public works idea, the chairman warned that such projects could lead to the "destruction of the Government's credit." Though the earlier WPA public works projects seemed to have the largest afterglow of all the many New Deal initiatives, the emerging coalition of southern Democrats and northern conservative Republicans now opposed reinvigorating any such public works agenda to employ returning soldiers.[36]

The resistance to new public works indicated just how expansive the reconversion debate in Congress became. It touched on how virtually every aspect of the economy would be managed after the war. The debate gave voice to an enlarged conservative alliance that resisted any return to what they characterized as heavy-handed New Deal command-and-control policies. Meanwhile, the liberal-labor coalition sought to build on the federal government's role in economic planning. Roosevelt hoped to navigate a path through this new postwar politics to realize a full employment economy. On July 1, 1944, the president signed the Contract Settlement Act of 1944, and on October 3, 1944, the Surplus Property Act. Like any good political compromise, each of these possessed elements of both the conservative and liberal approaches to reconversion. The Contract Act created a new Office of Contract Settlement with rules for unwinding contracts and a dispute-resolution mechanism. While it gave the administration ample authority and flexibility, it was under terms the Congress defined.

As a nod to the liberal approach, the act protected small business by having the federal government ensure that the myriad small subcontractors be guaranteed adequate payment as contracts were dissolved. Even more reflective of the liberal approach was the Surplus Property Act, which benefited small business by reducing industry concentration as the government disposed of plants and properties it owned during the war. Even the ship magnate Kaiser, who had told the *New York Times* panel that business should be largely left to manage the reconversion effort, became a principal beneficiary of government engagement. The favorable terms of reconversion allowed the firm to acquire aluminum facilities to compete against "the monopolist" Alcoa. And Kaiser's acquisition of Ford's Willow Run plant eventually allowed him to compete in the automobile industry.[37]

The reconversion debate was a window on the contours of a new postwar politics. It revealed how the post–New Deal conservatives were now eager to offer alternative visions of the government's future role in enabling technological development to stimulate economic growth. Meanwhile, the ever-evolving liberal-labor coalition, now finding itself losing its fragile hold on southern Democratic support, and with organized labor's reputation suffering, continued to promote government's central role in economic planning. Most importantly for the debate over full employment, both sides accepted the premise that the New Deal and the war mobilization fundamentally changed the relationship of the federal government to the economy. Conservatives now hoped to sustain the wartime model of a public-private partnership dedicated to growing defense-oriented technologies. Meanwhile the liberal-labor coalition, acknowledging the government's new role in enabling technological development, realized that if the government was capable of planning economic conversion and then reconversion, certainly it could help forecast future trends in mechanization and computerization that would allow for continued full employment.

With Senator Murray's full employment bill, the liberal-labor coalition thus hoped to build on the momentum that had begun when Robert Wagner took to the Senate floor in March 1928 to offer three pieces of legislation that addressed the realities of unemployment. Since that defining moment, the coalition was responsible for the enactment of a solid foundation of laws that helped workers realize their fair share of the benefits that companies reaped from ever-improving technology, and that provided some relief to those workers the new machines displaced. Yet as the GIs came home, no one could assure that the economy would not revert to Depression-era levels of unemployment. Thus, the liberal-labor coalition was intent on codifying a federal economic planning mechanism that was at odds with conservative expectations of the federal role in a new political economy—one that enabled rather than controlled. In this changing political climate of the postwar United States, the liberal-labor coalition would struggle to recapture its prewar political momentum on behalf of the worker. The reconversion debate had served as a harbinger of what was to come in the coalition's effort to enact full employment legislation, a debate that would test their vision against that of a formidable new conservative insurgency.

CHAPTER 4

Full Employment

The tides of science and technology cannot be held back;
our institutions must grow with them or be engulfed.

—Oscar M. Ruebhausen and Robert B. Von Mehren,
Harvard Law Review, June 1953

The week after Senator Wagner gaveled in his first day of protracted hearings on the full employment legislation he sponsored, the mystery that surrounded the secretive "Manhattan Project" vanished in a dramatic demonstration of atomic energy's destructive force. The atomic bomb that exploded over Hiroshima on August 6, 1945, revealed just how rapidly technology could advance beyond political grasp. In the moment there was, of course, euphoria over a war ended. Officials were then quick to reassure the public at large that such a stunning new destructive force could be harnessed for peaceful uses. The *New York Times* ran a series of articles on the promise of the new technological age, declaring, "For better or for worse we have entered a new era in the history of mankind." As part of the sudden paradigm shift, with technology now the panacea rather than the problem, the liberal-labor coalition followed suit, embracing technology as the means of sustaining the momentum of wartime full employment.[1]

Having already gained credence during the reconversion debates, the new political rhetoric that embraced the possibilities technology held for job creation now took full flight. In the postwar era, techno-optimism was widely shared. Everywhere, it seemed, there was an almost heady excitement over how the force of technological advancement that spurred victory could now be turned to extend prosperity. The so-called "First Lady of American Journalism," Dorothy Thompson, exuded the new spirit. Writing in her popular syndicated column, she exclaimed, "Technology in the service of a completely converted and purified world spirit can lift mankind to a material level where old antagonisms become irrelevant, explanations recognized as irrational, and a material civilization and deepened spiritual culture flourish in a reborn world." It was as if a world embodying FDR's 1941 Four Freedoms speech were suddenly realizable. In the glow of Axis surrender, other gleaming prognoses poured forth. The Yale University Labor and Management Center predicted that productivity would only continue to increase over the next five years. The war, it noted, had greatly accelerated research into new industrial processes that once would have taken decades to develop. The dividends of the investment in the nation's much-expanded university and industrial research system would now manifest in the form of myriad new products and conveniences. In a series of articles on the nation's impressive research capabilities, the *Christian Science Monitor* writer Harold Fleming observed that although pure research atrophied during the war in favor of applied research, the United States' investment in R&D was now greater than ever before: "At last the story can be told of the breath-taking new devices and products which the United States public is about to be able to buy as a result of the enormous technological advances of the last few years."[2]

Indeed, the US war production machine quickly transformed into a postwar economy scarcely imaginable before the war. A special congressional subcommittee on demobilization counted 839 new gadgets, products, materials, processes, inventions, and techniques attributable to wartime technology. Whether the subcommittee's methodology was sound or its count accurate, the technology that helped win the war seemed poised to deliver unprecedented peacetime prosperity. World War II–derived technologies, including the use of nuclear power for electricity, commercial uses of advanced radar systems, new lifesaving pharmaceuticals (e.g., penicillin and quinacrine), commercial air transport systems, and synthetic fibers substituting for scarce rubber (rayon), were among the obvious early transfers finding their homes in the postwar economy. As the *1945 Collier's Yearbook* proudly asserted, "In the fields of science, particularly in physics, aviation, chemistry, engineering and related subjects, enormous progress is being made as a result of the war's demands."[3]

Seizing on the pervasive techno-optimism, many business leaders dismissed a return to the prewar economy, where limited production coupled with increased productivity led to reduced and stagnant employment. Envisioning how the continuance of the cooperative business-government model might propel a new growth-oriented economy, a progressive cadre of businessmen began to encourage their own version of economic planning. For example, War Production Board member and General Electric CEO Charles E. Wilson (Charles Edward Wilson, as distinguished from his contemporary, Charles Erwin Wilson, the CEO of General Motors), having witnessed the results of the dynamic business-government partnership during the war, urged that the postwar government continue to infuse money into research to spark the next wave of technological advancement. Wilson, like other progressive business leaders, accepted the popular Keynesian emphasis on government fiscal and monetary policies as essential to raising the capital necessary to spur technology's salutary role in increasing productivity. Unlike their Depression-era corporate counterparts, Wilson and this new group of business statesmen sought an energetic and expansive postwar economy built on notions of abundance rather than scarcity. But although business leaders and their conservative congressional allies endorsed the broad concept of full employment as part of the postwar economic order, the ensuing debate revealed that they differed significantly from the liberal-labor coalition on how to achieve it.[4]

The liberal-labor coalition maintained its support for a strong social safety net and the need for public works projects as fail-safes in anticipation of sudden economic downturns. At the same time, a post-Depression new liberalism sought to sustain the momentum of the New Deal's stimulative policies while also acknowledging technology's potential to sustain near-wartime employment levels. "Full employment" legislation thus became an early political test. Would the conservative champions of technological progress recognize that guaranteeing full employment should be the price of a reconstructed social contract? Organized labor, still the central player in the liberal-labor coalition, aggressively continued to bargain for new contracts that unlocked its share of productivity gains. At the same time, however, labor leadership also enthusiastically joined its liberal allies in the political arena in order to promote the idea that the federal government must put in place a central planning mechanism that could foresee and accommodate any future step changes in technology that were likely to cost jobs. The liberal-labor coalition writ large was willing to share the vision of those in the business community who viewed technology as central to a productive postwar economy that would expand the blue-collar middle class. But it expected more from conservatives than just conciliatory rhetoric about the technological future. The liberal-labor coalition now sought legislative guarantees that full employment was the enforceable norm, and sought to institutionalize a planning mechanism that would guarantee full employment.[5]

Sensing the shifting political sands of post war politics, Walter Reuther, still a year away from being named head of the United Auto Workers (UAW), emerged as the fresh face of organized labor. In a lengthy op-ed in the *New York Times*, he focused on the challenges of the transition ahead. "There can be no return to the balmy Palm Beach climate of the nineteen twenties," the thirty-eight-year-old UAW vice president wrote. In making his case for "full production, full employment, and full distribution," Reuther focused on the gap between the nation's technological sophistication and its ability to fairly distribute technology's benefits. He put forward a new plan for keeping the US wartime industrial juggernaut running at full tilt. As he succinctly stated, "The will to change corporate habits and to renounce a profitable business simply wasn't there." Reuther thought it imperative that laggard industries like rail and construction embrace technological advancement by modernizing their businesses. He recommended new innovative policy ideas, such as conferring public utility status on the construction industry, in order to tackle the growing problems of America's inner cities. For Reuther's energetic mind, the possibilities that emerged from a technologically sophisticated economy stretched the imagination. A month after the armistice, he put forward a vision of a vibrant technological future that would serve all regions of the country and raise living standards for all Americans. Reuther astutely measured what the potential new technological future held for the blue-collar worker, while acknowledging industry's newly anointed status within the halls of government. Politically, he well understood the value of applauding the abundance made possible by technological advancement. At the same time, he cautioned that industry should not be allowed to retreat from the bold vision he outlined of an "economic democracy within the framework of a political democracy."[6]

Prelude to the Debate

The shared principles that eventually would underpin the negotiated truce between labor and business were evident in these early postwar pronouncements by the UAW's Reuther and GE's Wilson. The seeds were being sown for what, by the end of the decade, became known as the Treaty of Detroit, a sweeping new contract between the UAW and the auto industry that would extend to other industries and set important precedents around the conditions of employment. But the prelude to this agreement was a strike wave affecting major industries and threatening the reconversion. Indeed, up until his death at Warm Springs on April 12, 1945, an understandably weary and ill president would continue to worry that the seething domestic labor and racial tensions evident during the war would boil over soon after it concluded. Though he would not live to see his concerns manifest themselves as reconversion began, Roosevelt's fears seemed well placed. Workers patriotically laboring fifty-two-hour weeks with no overtime suddenly saw their hours drop and their pay diminish. With the war now over, they considered their earlier no-strike pledges moot and any protections afforded employers from the wartime Smith-Connally Act no longer enforceable. Workers became understandably more resolute in demanding that their employers restore the eight-hour day and that wages go up. When they failed to see this restoration, it fueled their determination to take collective action. The result was some 4,750 work stoppages in 1945 alone, involving 12 percent of the nation's workforce. On January 21, 1946, 750,000 steelworkers began a nationwide strike, and Reuther himself led the UAW to strike against General Motors in late November. That same month, GE's Wilson rejected wage and hour demands by GE electrical workers, and they, too, walked out. Even after major companies settled with the unions, hundreds of wildcat strikes among suppliers prevented the return to full production. In total, the BLS reported an even higher number of work stoppages during 1946 (4,975), and defined thirty-one large work stoppages as involving ten thousand workers or more.[7]

The wave of strikes in a nation already exhausted by four years of war tried the patience of even traditional union sympathizers. When the United Mine Workers' John L. Lewis led coal miners in a nationwide coal strike beginning on April 1, 1946, President Harry Truman tried to dust off the Smith-Connally Act, declaring its emergency requirements still in place. The Supreme Court surprisingly backed Truman's decision. When rail operators threatened a strike on the heels of the coal debacle, Truman again reacted forcefully, threatening a government takeover of the railways. The combination of the rail and coal actions seemed the final inflection point. Public sentiment toward unions, already souring during the war, now seemed to be decidedly moving further against their seemingly callous exercise of what the *New York Times* called their "monopoly powers." This growing antiunion public sentiment bolstered those conservatives in Congress intent on weakening the FLSA by putting a permanent limit on strike actions.[8]

Moreover, as Roosevelt had feared, there was evidence of growing anxiousness over job security. Millions of GIs were returning home looking for work. Yet, throughout the war years, as the last wave of the Great Migration continued with the flight of African Americans from the rural South to the industrial North, black workers rightfully sought their place in the postwar US industrial economy. Perhaps less dramatic than the migration that followed World War I, what started slowly soon became a virtual rite of passage, particularly for younger black men. There was another black migration as well: some one million African American men and women who had proved their undying allegiance to the country as enlistees now returned home hoping to share in the benefits of a postwar economy. But the 1944 GI Bill contained qualifying criteria that effectively excluded minorities from its educational, housing, and health benefits. Moreover, upon their move northward, African Americans met with repeated instances of white backlash. In Philadelphia and Chicago, there were ugly incidents of racial violence pitting white workers against blacks seeking jobs. Meanwhile the assertion of white supremacy continued in the South, with brutal white riots against African Americans in Columbia, Tennessee, and Athens, Alabama. On the West Coast there were also incidents of white aggression against Asians, including Asian American veterans. The episodes harked back to FDR's earlier warning. Establishment elites began to speculate whether, with the victorious end to the war, the United States might now be experiencing the beginning of its own domestic unraveling.[9]

At the same time, and in spite of this backdrop of social unrest, the view that the social contract needed to expand to ensure jobs for all those willing to work began to gain its own momentum. A chorus of supporters joined President Truman in his full-throated support for the idea that technology might be just the elixir that would make the right to a job a reality, as articulated in FDR's Second Bill of Rights. Among the influential full employment proponents, the celebrated liberal columnist Walter Lippmann reminded his readers that government could now take action to diminish once and for all the specter of unemployment. Lippmann remained steadfast in his belief that full employment was not simply a domestic imperative, but an international one. Writing in his widely syndicated column, he maintained that "if the American economy, which is such an immense factor in the world economy, is kept working steadily at reasonably full capacity it will set up a demand for goods which will contribute enormously to prosperity almost everywhere else." Reflecting his progressive moorings, Lippmann believed that public policy could be just as innovative as science. In the wake of remarkable technological advances, Lippmann urged politicians to be as imaginative in anticipating technology's darker impacts by expanding their own horizons.[10]

The very term "full employment" was an increasingly popular and politically appealing trope. But what did it actually mean? And how would policymakers be satisfied it was achieved? The concept itself was the intellectual progeny of an English economist, William H. Beveridge, an adviser to the British government, which faced postwar challenges more severe than those in the United States. Beveridge freely acknowledged that the term was not to be taken literally. It was impossible, he said, to expect that "every man and woman in the country who is fit and free for work is employed productively on every working day of his or her working life." Rather, he defined full employment as a state in which there were always more jobs available than there were unemployed individuals. He foreshadowed the vexing challenge for policymakers in the postwar era. Unemployment was inevitable, particularly as technology in the workplace advanced and, as it had done prior to the war, substituted for work previously done by humans. What proponents of full employment legislation sought was a system that compressed the time lag between losing a job and finding another. Achieving such required more planning, better anticipation, and improved coordination between public and private sectors than had existed prior to the war.[11]

Beveridge defined full employment for England to mean that only 3 percent were unemployed at any one time. He considered this percentage acceptable because of what he termed "industrial friction." In modern economies, and given the pace with which new machinery was transforming work, there would always be a mismatch between the jobs available and the workers seeking jobs. As one prominent economist, William Haber, noted in 1944, "Full employment does not mean a job for everyone now included in the war-expanded labor force; nor does it mean absence of unemployment. In a dynamic society such as ours, with a labor force of nearly 60 million under peacetime conditions, it is possible to have some 3 million workers unemployed even under conditions of full employment. Full employment does mean, however, a strong sustained demand for labor; an opportunity for productive work at all times." In the United States, the legislative debate seemed comfortable with the overall framework Beveridge outlined. As Harold Fleming, the prominent business writer, wrote in 1944, "Full employment, whatever it may mean, has become the accepted creed of both political parties and for a wide range of groups in the community from the CIO to the NAM [National Association of Manufacturers]." Thus, not surprisingly, the ensuing postwar congressional debate focused less on what the term meant than on whether government needed to intervene to help achieve it.[12]

The Debate Begins

When Senator Wagner gaveled to order the Senate Banking Committee hearing on the Full Employment Act the week before Hiroshima, the now veteran liberal-labor-coalition stalwart reminded his colleagues that although the economy was still on full war footing, it was just five years removed from a decade-long Depression. Now serving his fourth term,

Wagner recalled how fifteen years before, he had stood on the Senate floor and boldly proclaimed, "The right to work is synonymous with the inalienable right to live. The right to work has never been surrendered and cannot be forfeited. Society was organized to enlarge the scope of that right and to increase the fruits of its exercise." In the postwar euphoria over the triumph of technology, Wagner thus reminded his colleagues on both sides of the aisle that full employment was as much responsible for victory as advanced technology.[13]

At the outset of these hearings, the liberal-labor coalition was encouraged by the fact that the bill's original sponsor, Senator James Murray, was joined by four progressive Republicans and three liberal Democrats (including Wagner) as cosponsors. Wagner told his colleagues that full employment should no longer be contingent on government funding of public works. Recognizing the need to begin to separate from traditional New Deal solutions, Wagner stated his conviction that Murray's bill "firmly rejects the proposition that public employment is the main avenue toward full employment." The key component of the bill was the proposed National Production and Employment Rights Budget, an annual planning mechanism that called on the president to annually aggregate the total amount of both public- and private-sector expenditures necessary to produce the gross national product at prices that would allow for full employment. The annual budget mandate would serve both as an analytic tool and as the basis for specific executive actions and policy directives. Though the legislative text was not specific, most assumed this critical function would be undertaken by the Bureau of the Budget (the precursor to today's Office of Management and Budget). The bill's authors thought that such a forecasting mechanism would provide the government with the information it required to predict forthcoming economic downturns and prescribe appropriate remedial steps well in advance of any impacts on employment levels. More sophisticated and detailed than the Federal Employment Stabilization Board that Wagner had offered in 1927, the Production and Employment Rights Budget marked the latest iteration of the liberal-labor coalition's continuing desire to better anticipate technology's next turn. If the economy was to be well planned, the authors argued, then such analysis must be conducted before, not after, economic downturns. As the centerpiece of the original bill, the elaborate budgeting device granted the president the authority to propose, and thus by extension enforce, a general program for private enterprise investment. In short, it provided government the capability to intervene in the market and steer private-sector investment sufficiently and prophylactically to maintain full employment.[14]

While Murray and Wagner assiduously avoided authorizing any new department or agency, the elaborate and comprehensive nature of the proposed Production and Employment Rights Budget seemed to beg for an NRA-like bureaucracy to administer it. This provision quickly became fodder for conservative critics. Conservatives were entirely comfortable with government's new partnership with the private sector, which increasingly funded the research and development of new technologies, enabling and empowering government's new corporate partners. It was far less appealing to these same conservatives when the government acted as a referee over private decisions regarding how capital might be spent. As proposed, the planning mechanism seemed to allow for the federal government to intrude directly into future business investment decisions. Even within the otherwise supportive Truman administration, doubts were raised early on as to whether the Bureau of the Budget was staffed properly to carry out such an extensive budgeting task, much less enforce its findings when necessary. For conservative critics, government planning, when placed in the context of national economic budgeting, was reminiscent of the heavy hand of New Deal bureaucratic power.[15]

From the perspective of Murray's ardent supporters, including, not surprisingly, what remained of the New Deal–era NRPB, the Full Employment Act represented the reinvigorated vision of an enlarged liberal state. Anticipating the likely conservative and business backlash to what would clearly be viewed as enhanced government intervention in the economy, other allies of Murray's legislation, such as the National Farmers Union and the Union for Democratic Action, joined the NRPB in acknowledging that technology was a vital contributor to their optimism over how a postwar economy could maintain full employment. Even the most forceful proponent of Murray's bill, James G. Patton, the president of the National Farmers Union, long a forceful critic of technology's role in decimating agricultural employment, modulated his passionate antitechnology bias. Patton had long witnessed how the mechanization of agriculture had resulted in the steady exodus of farmhands to the cities since the turn of the century. In its assessment *Technology on the Farm*, the Department of Agriculture noted that as a proportion of the workforce, US agriculture had declined from one-half to one-fifth by 1930. Patton observed that "in our highly industrialized economy, full employment is the farm relief measure of our time." So identified was Patton with full employment legislation that Murray's bill was often referred to as "the Patton amendment." When he testified before Wagner's committee, Patton acknowledged that "we are confronted with a bewildering array of discoveries, of devices and processes and techniques that can bring us an abundance never before dreamed of. Or equally our cleverness with material elements of our universe can bring untold human suffering." Conforming to the new liberal-labor orthodoxy, however, Patton acknowledged the potential for technology to allow for full employment in peacetime as it had done during the war. Patton would portray the Full Employment Act as a means for improving the standard of living for that growing number of farmers who could not afford to remain in agriculture. He saw the legislation as necessary for providing new job prospects for those farmworkers quickly being eliminated due to increased farm productivity. Rather than challenge technology, Patton used the opportunity before Wagner's committee to become part of the liberal-labor chorus seeking to reap technology's promise.[16]

Similarly, CIO president Philip Murray's testimony before Wagner's committee reinforced this general sentiment. "Our vision," he said, is that now, for the first time in our history, all Americans can have abundance because we can all work together with science and technology to create abundance." Murray thus joined Green and other union colleagues who came forward to affirm the view that technological progress could lead to rising wages and reduced prices. Murray in fact tried to co-opt the opposition even further, by expressing his hope that postwar conditions, aided by the proposed full employment legislation, would lead to closer coordination between business and government than occurred during the war. In short, the liberal-labor coalition and its parade of witnesses before Wagner's committee were doing what they could to extend an olive branch to their conservative skeptics.[17]

No Easy Path

Despite such overtures, however, the debate over Senator Murray's bill would split along familiar lines, prove unusually contentious, and mark a turning point in the trajectory of liberal-labor reform efforts. The Democrats still controlled the Congress. Though they had lost forty-three seats in the 1942 midterms, they had won back twenty-two seats in 1944, thanks to the coattails from FDR's decisive presidential victory. But the majorities masked the disaffection of southern Democrats, the increasing appeal of the new conservatism within Republican ranks, and eventually the political impact of the transition from FDR to Truman. The reconversion debate already hinted at what the liberal-labor coalition had in store. Following FDR's death and in the new postwar political milieu, Senate minority leader Kenneth S. Wherry (R-NE) defiantly announced, "The New Deal and this Administration [Truman] is having its wings clipped and from now on you can expect Congress to continue the clipping."[18]

In anticipation of a more formidable conservative challenge, the liberal-labor coalition sought to enlist fresh new allies to its cause. Paul A. Samuelson, who was on the verge of publishing *The Foundations of Economic Analysis*, a book that would send him on a trajectory toward the Nobel Prize, tried to argue that Murray's full employment bill was in the conservatives' best interests. In September 1945, the young MIT professor cautioned conservative defenders of a "free market" that "it is the well-wishers of the capitalist system who have by far the greatest stake in the maintenance of a high level of jobs and of effective demand." He argued that a modicum of planning simply continued the cooperation between business and government that had been so successful during the war. Business simply did not have the skills necessary to forecast large economic trends and should want government's assistance to make sure the economy was humming at full capacity. "A joint effort rather than mutually exclusive responsibility" had become the hallmark of the nation's mixed economy, he suggested.[19]

Yet such pleas were not enough to persuade opponents. To ardent conservative critics like Senator Robert Taft (R-OH), the jobs budgeting mechanisms in the bill raised the specter of government planning. Throughout the debate, conservatives reflected a fear that the legislation was a throwback to New Deal interventionism. Their rhetoric foreshadowed the Republicans' 1946 off-year election slogan, "Had Enough?" Just as they had during the congressional deliberation on reconversion, conservative Republicans found it advantageous during the full employment debate to join with the conservative southern Democrats who largely controlled the House leadership. These combined conservative forces, in their effective resistance to any provisions that smacked of centralized economic planning, in turn caused the liberal-labor coalition to recruit one of their own southern Democratic loyalists to lead the House effort. Wright Patman (D-TX) represented the northeastern rural counties of Texas and seldom retreated from his long-held New Deal principles. His own politics navigated between populism and liberalism. As the first chairman of the Small Business Committee, he spearheaded legislation that protected small mom-and-pop stores from the predatory practices of larger retail chains. First elected to Congress in 1928, Patman would become a staunch ally of the Roosevelt administration and committed himself to ensuring that the worker was protected in the postwar United States, and that his signature Robinson-Patman bill (1936), which curbed price discrimination against small enterprise, was an important piece of New Deal legislation. As he began hearings on his own counterpart to Murray's full employment bill, he cautioned his colleagues, "Let us not reconcile ourselves to a cycle of booms and depression."[20]

Though the new House Speaker, Sam Rayburn (D-TX), shared Patman's small-town Texas roots, on the issue of full employment legislation the Speaker genuflected to those southern Democrats who stood in opposition to Patman's bill. Despite Congressman George Outland's (D-CA) enthusiastic efforts in garnering some one hundred cosponsors for Patman's bill, some unfortunate wording in the legislation allowed the opposition to cleverly call for its joint referral to the House Committee on Expenditures, chaired by its cranky segregationist and conservative chairman, Carter Manasco (D-AL). Despite Truman's unflagging support and an extensive grassroots network of all the familiar New Deal constituencies, Patman's bill had the support of only six of twenty-one members of Manasco's committee. The chairman and his allies not only took their time moving the bill, but ultimately sent it to the House floor stripped of its key provisions.[21]

During extended hearings in his committee, Manasco and his conservative allies peppered witnesses with questions that betrayed their skepticism. When Millard D. Brown of the Philadelphia Textile Manufacturers Association testified that the bill would cause the complete government regimentation of the nation's economic life, Manasco sympathetically agreed by noting that there was no way short of a mandatory central planning mechanism to compel people to buy enough to propel the economy. The chairman then further led the witness to confirm the specter of a federally controlled economy. He rhetorically mused, "I am just wondering if our people are sufficiently alarmed with the essential dangers of a planned economy."[22]

With the committee so stacked with opponents of the legislation, Patman himself could complete only a minute of his opening testimony before the irascible conservative Republican congressman Clare Hoffman (R-MI) quickly interrupted to ask, "Did you say something about affecting taxes?" Another member then sought recognition to complain that there were some four hundred thousand men who simply did not want to work, while others who wanted to work were locked out because of strikes. Other members opposed to Patman's bill then joined in, claiming the bill was recreating the WPA and would in fact lead to slower economic growth. Badgered by committee conservatives, Patman ended up speaking more about what his bill did not do than what it did. He reassured committee members that the bill did not allow the government to operate industries, assign workers to jobs, guarantee prices, set wages, dictate production quotas, or allow for disclosure of trade secrets. Yet the parade of witnesses who would follow Patman over the House committee's twenty-one days of meandering hearings stretching over six weeks, purposely extended to delay the bill's progress, emphasized all these concerns to the delight of the majority of committee members, who were wary of a New Deal renaissance.[23]

In contrast to the resistance Patman faced, Wagner's Senate path was far less contentious. All but four of the sixty-seven witnesses in Wagner's Banking Committee hearings testified in favor of the bill. In contrast to the House, the bill was referred to no other committee, the hearings that began in July were concluded in ten days, the bill was reported by mid-September, and it passed by month's end. Though the debate on the Senate floor was contentious, even Senator Taft, the lead Republican opponent, voted for the bill after he successfully added an amendment to require that any expenditure for public works trigger new taxes (a 1946 version of the "pay as you go" provisions of later twenty-first-century budget debates). Yet even in seeking a bipartisan compromise, Taft was willing to see such taxes spread out over time, "so as not to require a balancing of the budget in any year, or any combination of years." Indeed, in the context of dire warnings about the prospects for rampant unemployment, Taft and some other conservatives were content to commit to the goal of full employment as long as unfettered Keynesian spending was not the solution.[24]

Perhaps most critical to the bill's overwhelming Senate passage was the role played by the former chairman of the Temporary National Economic Committee, Joseph O'Mahoney. No one seemed better prepared for the Senate floor debate. And although the Wyoming senator arrived late to the floor, he stood near the rear of the Senate on the afternoon of September 27 and repeatedly fended off the bill's critics. For those that feared Murray's bill would unleash a wave of government spending and control over the economy, O'Mahoney argued that it was designed to do just the opposite. If the budgeting and planning mechanism in Murray's bill had been in place during the Hoover era, O'Mahoney argued, the nation would never have spent the billions it did on public works and other New Deal measures. Having chaired the Depression-era committee that gauged the impacts of technology on unemployment, he was less inclined to embrace technology's promise, and he reminded his colleagues of the deleterious impacts of technology on employment. Raising again the specter of technological unemployment that he understood so well, he said, "The great danger is that with improving technology, the output of goods is so much greater than formerly that we are producing more with less labor than at any time in our history." O'Mahoney went to the heart of conservative opposition by repeatedly championing the measure's protection of small business. The country, he said, should not leave economic planning in the hands of those managing the largest corporations. "This is an American bill to make free enterprise work," he concluded, imploring his colleagues to vote in its favor. The final 72–10 vote in the Senate that O'Mahoney helped secure would represent the bill's high-water mark.[25]

In contrast to the Senate's quick march to passage, the less enthusiastic House leadership delayed floor action until December 1945. And when the bill did pass, it was in a much weakened version that stripped the bill's declaration of the right to employment, jettisoned its budget planning mechanism, and eliminated the ability of the government to require any tax or fiscal measures that might ameliorate economic downturns. Manasco's committee had deliberately dawdled through twenty-one days of hearings with some forty-five mostly skeptical witnesses who opposed Patman's legislation. Then Manasco himself led a small group of conservatives in drafting a weakened substitute that was reported from committee. Going into the election year of 1946, with conservatives waging a separate fight over the authority of the newly formed Price Control Board, the battle lines were drawn for the floor fight over Manasco's committee's substitute for Patman's bill.[26]

For his part, President Truman became an outspoken proponent of the original full employment legislation. Sensing the stalwart opposition that faced the Patman bill in the House, he used his September 6, 1945, speech to Congress on the postwar economic reconversion to call for the bill's adoption by the House. Truman reasserted that every US citizen willing and able to work ought to be able to have a job, and if the market failed to provide enough jobs, then it was "the ultimate duty of Government to use its own resources if all other methods should fail to prevent prolonged unemployment." Yet even the president's enthusiastic support could not change the vote count in the House. Manasco and his allies on his Expenditures Committee had reported Patman's bill unfavorably, with only three members voting for it. And when the legislation came to the floor, Manasco offered a substitute that one member derisively remarked "was written by the best minds of the eighteenth century." Congressman Jerry Voorhis (D-CA), who a young Republican named Richard Nixon would defeat in November, rising to defend the original Patman bill, suggested that the proponents of "free enterprise" were forgetful of how recently capitalism needed to be rescued from total collapse. These very same people who would go on to criticize every New Deal measure were the very ones, Voorhis argued, who "came and begged the Government of the United States to save them from ruin." Despite Patman and his allies' best efforts, the conservatives prevailed with their substitute by an overwhelming 255–126.[27]

Manasco and his southern Democratic allies were relentless in the House-Senate conference, which lasted only ten days. Wagner, reading the political tea leaves, largely left it to a key staff member to serve as his surrogate during the conference. From the outset, the House conferees refused to accept the term "full employment" or, for that matter, any language that hinted at government spending or the guarantee of full employment. Given the tenacity of the conservative attack, the outnumbered liberal-labor conferees realized that their national budgeting mechanism was dead on arrival. Holed up in Senate majority leader Alvin Barkley's (D-KY) office, the conferees produced a final conference report that was almost unrecognizable from the Senate-passed version. The best that Senate conferees could accomplish was to salvage the term "full employment" in the final version. With the commencement of the midterm election year, and with none of the conferees wanting responsibility for killing the legislation entirely, the final declaration of policy included the phrase "maximum employment." But it was at best a pyrrhic victory of sorts, and the final bill would be called the Employment Act of 1946.[28]

During the Senate's perfunctory debate on final passage of the conference report, Senator Murray, though conceding that important elements of the original bill were eliminated, reminded his Senate colleagues why he authored a more robust version in the first place. Businesses, he told them, simply could not be relied on to maintain continuous employment without government engagement. He noted how the 1930s was actually a period in which technology greatly increased productivity, yet no attempt had been made to maintain a general purchasing power in the hands of the people. Though the battle was lost, Murray was not retreating from his views on the need for a federal commitment to full employment. In the emerging field of public polling, an Elmo Roper poll published in *Fortune* the previous year revealed that 55 percent of those questioned believed that government and industry should work together toward full employment. Murray thus remained confident that his view was not only substantiated by recent historical experience, but also buoyed by public opinion. Yet Murray's artful floor speech could not disguise his disappointment with the final bill, a disappointment shared by Truman. The typically jaunty president seemed less than enthusiastic when he met a group of senators and congressmen who gathered for the White House signing ceremony on February 20, 1946. Speaking frankly, as was his custom, the president said the employment bill was "not all I had hoped for."[29]

Conservatives, by contrast, sensed that their victory marked an important turning point. Defending the conference report during the House floor debate on February 6, 1946, Manasco stated that the government should no longer take on the cost of providing jobs. "We [the House conferees] took the position our Federal Treasury, now in debt 279 billion dollars, could not undertake such a commitment," he told his House colleagues. Conservatives had continually said they were committed to the goal of full employment throughout each stage of the bill's odyssey. In the end, however, their words rang hollow. They had achieved an important victory by successfully eliminating the National Production and Employment Rights Budget that allowed the president to proactively take measures that would stimulate the creation of jobs when an economic downturn was forecast to sustain full employment. This was the essence of the liberal-labor coalition's argument. The conservative *Chicago Tribune* praised the House conferees for rejecting the idea that the government should ultimately take responsibility for ensuring full employment. The more sympathetic *Washington Post* disappointedly acknowledged that the Manasco substitute was "a dud."[30]

The debate over full employment legislation had begun with liberals embracing the promise of technology's continued momentum in the postwar era while maintaining their insistence on a strong federal economic planning function. It ended with conservatives asserting their newfound influence by limiting what government could do to ensure jobs for all who sought them. In its initial stages, the bill was referred to as a fitting remembrance to President Roosevelt, his Four Freedoms speech, and his Second Bill of Rights. For Roosevelt, "The first obligation of government [was] the protection and well-being, indeed the very existence of its citizens." Nonetheless, for the new conservative insurgency, the original legislation overstepped. In its compromised final state, the new law reflected the growing power of a new conservatism that wanted to change the balance of government leverage over the economy. As signed into law, the Employment Act of 1946 retained the commitment to full employment as a matter of national policy, yet it had no enforceable means to achieve it. The new law simply required the president to establish a new National Council of Economic Advisers (NCEA) and deliver an annual economic report to a newly created congressional joint committee that was empowered to review and oversee the report's findings. But the legislation did not expand the government's authority to take any measures that might mitigate future economic downturns or stimulate greater hiring. Absent were provisions that would spur greater private-sector investment in order to stimulate job growth. Conservative trepidation over more government spending and an expanded bureaucracy to plan and execute a "federal jobs budget" prevailed over liberal desires to incentivize the continuance of cooperation between government and business in planning and investing for the economic future and avoid a return to a cycle of boom and bust.[31]

By July, Truman nominated the first NCEA chairman as the new law required. After a perfunctory confirmation process, Edwin G. Nourse, a vice president of the Brookings Institute and a familiar face to those committed to full employment, was confirmed by the Senate. Just two years before, the tall, thin, bespectacled economist, known for his work in agricultural economics, came before Senator Murray's Subcommittee on Military Affairs to present a compelling argument for the inherent soundness of the US free enterprise system. He concluded that what had occurred during the Depression was a failure of US business to compete at full throttle, and he embraced the widely accepted Keynesian approach to stimulating a technologically sophisticated, growth-oriented modern economy. Nourse would later side with Murray on the belief that the country needed a strong economic planning capability. Yet at his confirmation hearing to become the first chairman of the NCEA, Nourse was sufficiently attuned to the political winds to sidestep a rehash of the full employment debate. Instead he acknowledged that private enterprise should be allowed to determine how best to meet what he euphemistically described as "the broad ends of national welfare." As he prepared to preside over what would now be the only government apparatus aimed at encouraging full employment, Nourse told an interviewer that he was comfortable with allowing the private enterprise system to rise to the occasion. When he later spoke to an industrial gathering sponsored by the National Association of Manufacturers, Nourse charged his business audience to actively cooperate with labor and agriculture to ensure that the cycles of boom and bust did not recur. He soberly reminded the executives that they bore significant responsibility to foster competition, keep prices reasonable, and maintain well-paying jobs. His attempt at persuasion was as far as his mandate allowed. "We have dropped panaceas," Nourse would say of his truncated authority. Even the most ardent advocate of economic planning realized that the liberal-labor coalition's vision for a federal economic planning authority would have to wait for another day.[32]

More of That Automatic Business

By the time Nourse's NCEA issued its first report in December 1946, there was no mention of the word "automation." Yet Nourse was hardly alone in failing to see the next major technological turn that would transform the industrial workplace. The word had only just been coined. Earlier that year,

D. S. Harder, Ford's vice president of manufacturing, toured the company's new engine manufacturing plant in Cleveland. Witnessing the benefits from recently installed automatic controls, he told plant personnel that what they needed was more of the sort of material handling equipment that would bring the company even greater productivity. "Give us more of that automatic business—that automation," Harder is reported to have said. Whether or not Harder coined the word "automation," it slowly made its way into industrial and then popular parlance. In 1948, Harold W. Sweatt, the president of Honeywell Corporation, suggested that "modern technology was pointing toward completely automatic operation of home and factory." The new concept of automation, however it was defined, seemed a marked departure in the technology of manufacturing. Previously, machines had eliminated mostly unskilled laborers. Now these new devices threatened to take out skilled workers as well.[33]

An early and disturbing vision of what the automated future might hold for workers appeared in the November 1946 issue of *Fortune* magazine. The authors of an article entitled "The Automatic Factory" were two Canadian radar experts conversant with the very technology that Wiener had helped to refine during the war. In their visionary piece, E. W. Leaver and J. J. Brown detailed just how a future automatic factory might run entirely on its own. Their description was accompanied by the artist Arthur Lidov's four-page conceptualization of a totally self-sufficient, automated factory. Thus, both in words and in pictures, *Fortune* conveyed a bold futuristic vision of a totally integrated, self-operating plant, where a master control unit with punch cards and perforated tape served as the electronic brain for the entire factory. In none of the illustrations is a single person in sight. "Nowhere is modern man more obsolete than on the factory production floor," the authors concluded. Leaver and Brown optimistically envisioned how this new world would force society to find a better use for men than to make them mechanical operators of machines.[34]

Fortune boldly positioned its vision of a futuristic factory as the next logical phase in the steady evolution of technology in the workplace. The antecedents for just such a totally automated industrial operation dated to the 1920s, when a range of increasingly sophisticated industrial control mechanisms, from pneumatic to electro-mechanical devices, were introduced to regulate parameters such as temperature, pressure, and flow rates. By the late 1920s, new sensors and actuators resulted in workstations that controlled industrial functions without human supervision. Industries that had previously segmented work into discrete batches could more seamlessly link them together in one continuous process. During the Depression, the momentum for automatically guiding a continuous production process took several steps forward, as pioneering scientists at emerging corporate giants like General Electric applied their new electronic networking theories. During World War II, Bush's scientists further perfected electronic monitoring devices that enhanced productivity by allowing for continuously calibrated, machine-driven adjustments throughout the manufacturing process. And by the end of the 1940s, the first generation of industrial computers were helping to operate entire facilities such the Port Arthur refinery in Texas and the Monsanto Ammonia plant in Louisiana.[35]

Understandably, however, the mention of the notion of automation in the president's annual economic report would not come until 1954, and then it was in the form of a passing reference to "an active demand for automatic controls," and "electronic automatisms." In December 1946, as Nourse was just beginning to staff his newly formed NCEA, he was hardly alone in not being able to forecast the extent to which the industrial workplace was about to be transformed yet again. He used the NCEA's brief inaugural report simply to summarize the Employment Act itself, reassuring readers that there was no intent that business be coerced or constrained into achieving full employment. Clearly, Nourse's tone conformed to the new political realities. Though the historic proponents of full employment legislation, like James Patton and Alvin Hansen, never advocated for 100 percent employment, their aim was to build on the now well-established precedent of government and business cooperation. Nourse's first report even acknowledged that Americans had come to expect, even demand, that the federal government play a more active role in the economy. And yet, as the report clearly indicated, the NCEA's role was limited to analyzing, forecasting, advising, and consulting—a far cry from the broader liberal-labor coalition's original vision.[36]

Meanwhile, as Nourse carefully navigated his new role, in the 1946 midterm elections the electorate signaled that perhaps they had had enough. Republicans suddenly found themselves the majority in both chambers of Congress for the first time since 1932. Buoyed by their newfound ability to thwart liberal-labor coalition goals while still in the minority, and helped by the reaction to organized labor's wartime activism, the conservative insurgency now turned its attention to scaling back labor's right to strike. Their new efforts at retrenchment, including the enactment of the Taft-Hartley amendments to the FLSA, put a halt to the trajectory of liberal-labor reforms of the previous two decades. Republicans again reasserted their faith in the market and were encouraged by an ever-expanding postwar economy. Even the Truman administration felt obliged to join in the chorus. The opening sentence of the NCEA's first real annual report on the state of the economy triumphantly proclaimed, "As the year 1947 opens, America has never been so strong or so prosperous. Nor have our prospects ever been brighter."[37]

No doubt, the postwar US economy seemed increasingly rosy and jobs plentiful. From the end of the war to 1950, consumer spending would increase some 60 percent. Unemployment remained under 4 percent. Emblematic of the pervasive optimism, a Kiwanis International pamphlet proudly displayed a patriotic red, white, and blue cover with the title *It's Fun to Live in America*. As earlier predicted by technology's champions, the scientific discoveries of the war were now transformed into products in a robust new consumer economy. The now publicly celebrated Vannevar Bush wrote, "Advances in science when put to practical use mean more jobs, higher wages, shorter hours, more abundant crops, more leisure for recreation, for study, for learning how to live without the deadening drudgery which has been the burden of the common man for ages past." And yet, while Bush was writing and testifying to science's enhanced role in the dynamic postwar economy, Norbert Wiener, whose *Cybernetics* was published in 1948, was busily writing letters and telegrams to anyone who would listen about what he viewed as the perils of the automated future about to burst forth in ways unimaginable and with untold consequences for factory workers.[38]

CHAPTER 5

Automation Arrives

And not one of them buys new Ford cars either.

—Walter Reuther, responding to a Ford executive at Brook Park, Ohio, who remarked that none of the automated machines paid union dues

"First, I should like to explain who I am," began Norbert Wiener's August 13, 1949, letter to the UAW president. The letter landed in Reuther's mailbox at the worst possible time. The new labor boss was executing his closing strategy at the end of the UAW's latest negotiations with Ford. The mega-negotiation would result in the Treaty of Detroit, the breakthrough agreement that laid the basis for the relationship between labor and management across industries over the ensuing decade. With his charismatic public persona and adroit internal maneuvering, Reuther had only recently ascended to the presidency of the UAW in March 1946. Despite tumultuous fractionalization within UAW ranks, and a dramatic attempt to assassinate him, he already had skillfully navigated earlier negotiations with GM (1945–46) that resulted in wage increases and controls on the prices of the cars it sold. In the months that followed, his reputation as a tough negotiator carried over to the internecine battles within the UAW ranks over whether members needed to sign the anti-Communist pledge required by the new Taft-Hartley Act enacted by the Republican Congress in 1947. While of course outraged by the law's limitations on job actions, he also well understood Cold War public sentiment. Thus, he boldly supported the law's anti-Communist platform, boosting his credibility with management by appearing fair minded. Deftly straddling boundaries both inside and outside the union halls, he engaged the 1949 Ford negotiations with a need to consolidate his still-fragile internal power. When Wiener's telegram arrived, Reuther's own position remained tenuous.[1]

In these stressful circumstances, it would have been perfectly understandable if Reuther had simply dismissed a lone letter from an eccentric MIT professor. Yet Wiener's passionate concern about an automated future caught the union leader's attention. The head of 919,200 UAW members immediately telegrammed a response to the anxious professor:

> DEEPLY INTERESTED IN YOUR LETTER. WOULD LIKE TO DISCUSS IT WITH YOU AT YOUR EARLIEST OPPORTUNITY FOLLOWING CONCLUSION OF OUR CURRENT NEGOTIATIONS WITH FORD MOTOR COMPANY. WILL YOU BE ABLE TO COME TO DETROIT? WALTER REUTHER, PRESDIENT UAW CIO. AUGUST 17, 1949.

It seems remarkable that Reuther responded at all. Just two days before Wiener's telegram arrived, the UAW's 106,000 hourly union workers at Ford voted to go on strike over the terms of the company's proposed contract. Reuther himself marched at the head of a walkout of some 64,000 employees at the massive River Rouge plant, where the company had cleverly "speeded up" the assembly line, adding yet another distracting tactic in an attempt to erode the union's negotiating leverage.[2]

Reuther neither embraced nor dismissed automation. Rather, in his statements and actions, he concentrated on negating automation's worst features through innovative bargaining tactics while advocating government measures that more effectively anticipated an automated future. He also seldom missed the opportunity to rhetorically caution the most ardent corporate proponents against their overly enthusiastic embrace of the automated workplace. Reuther's dilemma was how to navigate the automated future while not appearing to be a reincarnated Luddite. His challenge was trying to reconcile workers' wages with the profits that industry derived from steady technological advance. Like the AFL's Green in the 1920s, Reuther would maintain a relentless focus on ensuring that his members got their fair share of the rewards from increased productivity.[3]

If the Taft-Hartley Act had weakened labor's ability to use strikes as leverage in its bargaining, it was not apparent from the agreement that Reuther, following on his earlier deal with Ford, in turn negotiated with GM in May 1950. The five-year agreement included the important recognition of cost-of-living adjustments and an annual improvement factor that substantially increased wages over the life of the contract. Analysts believed the agreement resulted in a 20 percent increase in the standard of living over the next half decade, thus at least partly capturing the productivity gains that the auto company had realized since its adoption of the new automatic equipment. The sweeping agreement, soon assented to by the other automakers and then by other industries, was widely viewed as an important victory for labor and further cemented Reuther's reputation. As Daniel Bell, then the labor editor at *Fortune*, commented, though GM may have succeeded in paying less than it should have, the contract threw overboard conventional wage theory as previously determined by political influence, which sustained the idea that labor provided but should not realize the surplus value it created. In addition to including health care benefits and establishing company contributions to a pension plan, the agreement also genuflected to an idea Reuther would continue to pursue in the halls of Congress: the beginnings of a guaranteed annual wage. Under the agreement, GM would contribute four cents an hour to worker wages to account for increases in productivity. "Such a wage is an important part of economic justice and is a necessity for full economic production," Reuther said as he acknowledged recognition of the concept. For the moment, technology and employment seemed to have found a basis for accommodation. But Reuther, never content, wasted little time suggesting that he would soon seek to expand the guaranteed annual wage.[4]

During these groundbreaking negotiations, Reuther had carved out time to fly to Boston to meet with Wiener. On March 14, 1950, the two sat across from one another at a hotel restaurant to discuss the threats posed by automation, and they each walked away impressed by the other's commitment. They agreed on a series of ideas as to how best to publicly convey, both domestically and globally, their agreed-on position about the potentially devastating impacts of automation. They even planned the establishment of a joint initiative between concerned scientists and labor officials, though the plan never came to fruition. Reuther thought so highly of the MIT scientist that he offered Wiener a speaking role at a 1952 UAW-CIO meeting. Though Wiener's clinical depression and increasingly ailing heart forced him to decline the invitation, he had accomplished what he had set out to do. In Reuther, he found a devoted partner in pressing the case against the human toll of automation, and Reuther soon found himself at the center of the national debate over policies to mitigate automation's effects.[5]

Their shared sense of urgency was not misplaced. In the early 1950s, automation made steady advances well before there was public, let alone political, recognition that technological unemployment had assumed yet another dimension. The dawn of the automated era occurred as the liberal-labor coalition's political effectiveness—indeed, its very coherence—was being challenged. The enactment of Taft-Hartley and the neutering of full employment legislation were indicative of the sudden reversal of fortune. The flattening trajectory of legislative reforms aimed at protecting workers, both those with and without work, occurred at a moment when Reuther and those in his union ranks were growing concerned with what appeared to be a dramatic step change in industrial operations. When the Ford Motor Company announced that it was opening a new six-cylinder-engine plant in Brook Park, Ohio, in 1951, it proudly announced that Ford would soon produce 4,600 new engines daily with 4,500 employees. By February 1953, Alfred Granakis, the president of the plant's UAW Local 1250, began sounding alarms. He reported to Reuther, "I am greatly distressed and deeply concerned about the problem of automation." Granakis's anxieties were not misplaced. By 1954, Ford added a another highly automated eight-cylinder-engine plant to the Brook Park complex. These combined operations became the hub of what the company proudly claimed as its "automation leadership." On inspecting the new plants firsthand, Reuther would describe one of the processes as being like one enormous, integrated machine tool, which took up some forty thousand square feet and involved forty-four automation units that effortlessly performed 530 cutting and drilling operations on each engine block casting. After Reuther toured the operations, one excited Ford executive proudly pointed to the line of machines and jocularly boasted that none of them paid any union dues. "And not one of them buys new Ford cars either," Reuther quickly responded, harking back to the basic principles of the company's founder.[6]

The Changed Political Landscape

As automated processes spread—upsetting the equilibrium of a newly reconverted economy—the nation's politics wrestled with its own shifting balance of power. In the first half of the 1950s, as congressional majorities swung back and forth between Republicans and Democrats, southern Democrats would continue to pose a legislative obstacle to major labor law reform. In fact, the best the liberal-labor coalition could do for the worker was to pursue modest increases in the minimum wage and unemployment compensation, while attempting to claw back some of the losses suffered to the Wagner Act. In this new, more conservative political milieu, organized labor, as Reuther's recent success demonstrated, found the bargaining table far friendlier than the political arena. Meanwhile, the mobilization during World War II, and then the successful reconversion, suggested to Republicans how government could play a role in enabling technological development and helping to grow new defense-related markets in the Cold War era. Even the most ardent conservatives could lay down their antigovernment predilections if the result was new defense contracts for their states and districts. When President Dwight D. Eisenhower gave his first State of the Union address, he was anything but the antigovernment crusader. "There is, in our affairs at home," he said, "a middle way between untrammeled freedom of the individual and the demands for the welfare of the whole Nation. This way must avoid government by bureaucracy as carefully as it regards neglect of the helpless." At the outset of his administration, Eisenhower championed government's continued support of technological research as essential to prevailing in the Cold War. Though at the end of his presidency he warned about the government's domination of university-based research and the military-industrial complex, over his eight years in office the government's research and development budget grew substantially, from $5.3 billion in 1953 (1.36 percent of GNP) to $13.7 billion in 1960 (2.60 percent of GNP).[7]

In part, the liberal-labor coalition had itself to blame for the way in which postwar research evolved. As early as 1942, Senator Harley Kilgore (D-WV), a longtime New Dealer, introduced the first of a series of bills to establish a National Science Foundation (NSF). Kilgore saw a linkage between scientific research and the achievement of social goals, including full employment. His approach sought funding for the social sciences and ran counter to an Office of Scientific Research and Development (OSRD) blueprint that Bush had authored, describing how the government role in science might best transition after the war. His detailed July 1945 memorandum, entitled "Science: The Endless Frontier," mapped out the establishment of an independent scientific research agency. Bush was pleased when a coalition of concerned scientists shaped the debate over how atomic energy should be managed in the postwar era. This resulted in the prompt enactment of the McMahon Act, which directed that the management of atomic energy for electrical power and other peaceful purposes be placed in civilian hands. He hoped a similar coalition of progressive scientists, led by the American Association of Scientific Workers, might have equal force in the debate over the formation of the NSF.[8]

Yet the contentious five-year fight over the shape of the NSF devolved into a debate over the organization's independence and accountability and pitted factions of the liberal-labor coalition against one another. Amid the bickering over the structure and scope of a peacetime, civilian scientific agency, the War Department took increasing ownership of federal research dollars. By the end of 1945, Bush found his own OSRD eliminated by Truman's Presidential Scientific Research Board. In what turned out to be a brief role Bush would play at the Pentagon, he witnessed firsthand how ongoing government-sponsored research was becoming the province of military rather than civilian planners. As the debate over the NSF became bogged down, a newly established Office of Naval Research in the Pentagon began syphoning the postwar research budgets previously controlled by Bush's OSRD, precisely the trend Bush feared would occur. As proposed in the Bush plan, the NSF was to be walled off from politics as much as possible to allow scientists to govern the nation's future research agenda. Yet now the military seemed as much in control of future federally funded research as it had been during the war. As Congress quibbled over the structure and responsibilities of the NSF, a steady flow of federal research money began to flow from the Defense Department to universities and private-sector contractors.[9]

This new postwar surge of government investment in defense-related research greatly benefited the growing Sun Belt region and placed it at the epicenter of an increasingly more conservative postwar political landscape. The region, which stretched from southern California through the heart of the old Confederacy, touted its increasingly favorable business climate as it now competed head-on with the industrialized North for new plant locations and corporate headquarters. The New Deal had tempered the southern wage advantages that had contributed to the last wave of the Great Migration. The South now offered a nonunionized, low-wage environment and mechanized new operations, nowhere better exemplified than in a growing textile industry that was crippling its older, less modernized New England competition.[10]

Both the old South and the newly industrialized Southwest benefited from New Deal programs that had measurably improved their infrastructure, including the availability of inexpensive water and electricity. Coupled with the arrival of air conditioning and the allure of Eisenhower's promised interstate highway system, the Sun Belt was rapidly becoming an attractive option for business expansion. Moreover, a ham-handed and failed postwar effort by the CIO to unionize the South, its so-called "Operation Dixie," served only to reinforce the attractiveness of these "right to work" states to northern businesses trying to escape rising labor costs. The aerospace and defense industries, buoyed by new government contracts, wasted no time in realizing the benefits of relocating. They supported a rising breed of successful businessmen turned politicians who formed the backbone of the new Sun Belt conservatism. When the Phoenix department store magnate Barry Goldwater announced he was running for the city council, he told his employees, "I don't think a man can live with himself when he asks others to do his dirty work for him." In addition to reducing government interference with the market, part of that dirty work for these new conservatives was keeping the Sun Belt free of organized labor.[11]

At the same time, since the enactment of the Wagner Act, the relationship between management and labor in the industrialized Midwest and Northeast became increasingly institutionalized. Companies beefed up their own personnel bureaucracies and retained a phalanx of consultants to anticipate new union demands. And though Taft-Hartley arguably made it more difficult to strike, it nonetheless reaffirmed the union's right of action and arguably strengthened labor's hand at the bargaining table— one they put to advantage, as the Treaty of Detroit demonstrated. Faced with the endless prospect of recurring and contentious union negotiations, companies increasingly looked toward the Sun Belt's improving business climate. Ironically, in a gesture of reassurance to northern industrialists, the head of the Tennessee Valley Authority (TVA), a well-recognized New Deal success, suggested that northern industry should have no fear about unions gaining inroads in the South in the future. The "fears its [organized labor's] shadowy shapes engender . . . will vanish," he proclaimed. His statement validated the southern right-to-work mantra that branded southern state boosterism.[12]

Freshman Senators

Indeed, as the pronouncement from the TVA indicated, in the changing politics of employment, the liberal-labor coalition seemed flat-footed and unable to counterpunch. During the brief recession of 1948–49, Senator James Murray, lead sponsor of the original Full Employment Act, offered the Economic Expansion Act to stimulate the ailing economy. He warned his colleagues that business was going back to the economics of scarcity rather than toward abundance. He recalled the not-so-distant Great Depression, and then in true Keynesian fashion proposed that a massive $15 billion be authorized to stimulate economic growth. His bill was replete with familiar provisions that bolstered unemployment insurance, funded public works projects, and ensured the retraining and relocation of displaced workers. Though he tried to placate conservatives by proposing that government aid would come in the form of recoverable loans, his bill failed to capture much support.[13]

What hope there might be for a renaissance of liberal-labor reform now rested with a new cohort of freshmen senators elected in 1948, which included the former economics professor Paul Douglas (D-IL) and the progressive Minneapolis mayor Hubert Humphrey (D-MN). Like Wagner during his own freshman term, these new liberals seemed willing to defy political odds to carry the torch of reform forward. Though they found themselves navigating a political context increasingly defined by anti-Communist sentiment, an emboldened conservative movement, and a more organized business community, they also found mentors in their own ranks. Not the least of these was Lyndon Johnson, who arrived in Washington in 1931 as the lone aide to a newly elected congressman from Texas. Johnson possessed deep populist roots that extended into the heart of the East Texas Hill Country where he was raised. Early in his own life, well before the rise of Sun Belt politics, Johnson witnessed widespread unemployment and abject poverty, including the bankruptcy of his own family farm. Elected to the House in his own right in 1936, he quickly became a staunch New Dealer and captured the attention of FDR, who appointed him head of the Texas National Youth Administration. After a visit to Texas in 1937, where he met the first-term congressman, FDR told his political confidant Tommy Corcoran, "I have just met the most remarkable young man. Help him with anything you can." After six terms in the House, Johnson won his own election to the Senate in the class of 1948. There he wasted no time moving toward a leadership position. Florida senator George Smathers would later describe being around LBJ as being in the presence of "a great and overpowering thunderstorm that consumed you as it closed in around you." In four short years he would be leading his party in the Senate.[14]

In his campaign for Senate leadership, Johnson immediately befriended Humphrey, his loquacious and hyperbolic freshman colleague. He viewed Humphrey as an invaluable interlocutor with the "liberal wing" of the Senate Democratic caucus. Humphrey's dramatic speech on a proposed civil rights platform plank at the Democratic convention in Philadelphia in July 1948 thrust him into the national spotlight. When he arrived to take his Senate seat in 1949, however, the former mayor found that his progressive politics were at odds with a Senate still largely dominated by a coterie of conservative southern Democrats with important committee chairmanships. In his early Senate years, Humphrey found himself snubbed by the conservative southern segregationist bloc, which included the likes of Richard Russell (D-GA) and John Eastland (D-MS). Yet it was his fellow freshman Johnson, with his strong relationships with these southern Senators, who reached out to Humphrey. Seeing the unabashed idealism of his colleague often in too-enthusiastic display, Johnson played the role of political mentor, never hesitating to give Humphrey frank advice about the realities of Senate politics in long conversations, usually over drinks in Johnson's Senate office. "If you liked politics," Humphrey later said, reflecting on his many long talks with Johnson, "it was like sitting at the feet of a giant."[15]

It did not take long for Johnson to realize his ambition of Senate leadership. Though he was still in his first Senate term, the 1952 elections presented him with an immediate opportunity. Eisenhower's landslide victory created long Republican coattails and swept out of office prominent Democrats, including their leader, Arizona senator Ernest McFarland, who lost to the rising conservative force, Barry Goldwater. With Democratic leadership up for grabs, Johnson seized the moment. He knew he would have little trouble enlisting his southern Democratic allies as his core base of support. Yet in order to defeat the inveterate New Dealer senator Murray to become the minority leader, he would have to broaden his base with liberals beyond Humphrey. Thus, in the wee hours of the early morning on November 5, he called to congratulate a victorious new congressman, John F. Kennedy (D-MA), who had just eked out a victory over the incumbent Republican, Henry Cabot Lodge, to become the junior senator from Massachusetts. Johnson instinctively realized that Kennedy was another liberal he might persuade to vote for him to become minority leader.[16]

With his penchant for making politics a game of addition, Johnson would win the minority leadership, becoming the Senate's youngest leader in either party. In his new role, he placed every member of his caucus on at least one significant committee, a practice that would become known as the "Johnson Rule." In Kennedy's case, Johnson overrode pressure from the young senator's father, who, already looking ahead, wanted his son to serve on Foreign Relations. Instead Kennedy was initially assigned the Government Operations Committee, which Johnson argued was critical in the context of the ongoing McCarthy hearings on un-American activities. More importantly from Kennedy's perspective, he also gained an assignment on the Labor Committee. This assignment made sense. Kennedy had served on the House Labor Committee, beginning with his first term in 1947, thus immediately receiving a baptism of fire in his opposition to Taft-Hartley. During Kennedy's brief three-term career in the House, the frail, almost gaunt-looking congressman kept labor issues in his otherwise distracted sights. The Senate assignment allowed Kennedy to solidify his reputation with organized labor, even as he badgered Teamster witnesses during the committee's hearings on corruption within organized labor. As the auto workers' general counsel Joseph Rauh reflected, "Whenever the UAW needed John Kennedy, he was there." Kennedy also realized that with the Senate Labor Committee assignment, he could readily pursue legislative initiatives aimed at bringing jobs back to New England, as he had promised during his campaign. He wasted no time in doing so.[17]

Depressed Areas

In the early evening of May 18, 1953, the Senate was in a period of morning business that allowed members the opportunity to speak about whatever they wished. Kennedy asked to be recognized so he could begin to make the first of three lengthy speeches on the state of the New England economy. For the next hour and a half, and for similarly extended periods the following day, and then again the next week, Kennedy held forth, offering an ambitious plan to help resuscitate the region's flagging economy. He was only four months into his first term, thus breaking with the convention that freshman senators refrain from speaking on the Senate floor in their first year in office. Kennedy, however, had decided that the crisis back home demanded he speak out, though out of respect he sought to remain somewhat inconspicuous by timing his remarks so that no other senators were competing for attention. His speeches reflected his concern about both the flight of industry from New England and the plight of the unemployed left behind.[18]

These speeches were a far cry from the soaring rhetoric that characterized Kennedy's later presidential speeches, even though the senator's young staff assistant Ted Sorensen, who would later write many of Kennedy's most famous lines, also composed these cut-and-dried tutorials on the causes and effects of the demise of the New England textile industry. Though much of the analysis came from work done by Kennedy's father's staff, the facts spoke for themselves. Since the war's end, some seventy separate Massachusetts textile mills migrated to the South or closed altogether. Prior to the Depression, the region was home to 80 percent of the industry; now it possessed only 20 percent. In his campaign for the Senate and subsequent visits home, Kennedy saw firsthand the devastation the manufacturing exodus brought to local communities. He concluded that they were too dependent on one or two industries, and the focus of his program was to have the federal government support job retraining, economic diversification, and factory modernization to help rectify the crisis.[19]

The case Kennedy made for the region aligned well with recommendations issued by the bipartisan New England Governors Association, as detailed in its own *Report on the New England Economy*. Kennedy's speeches spoke to his region's distress while artfully avoiding antagonizing constituencies that would be important were he to run for the presidency. He adroitly appealed to business by criticizing what he called the "shapeless mass" of

federal transportation regulations and arguing for a tax code that incentivized capital investment in new equipment rather than rewarding those who
profited from business liquidations. He further ingratiated himself with
organized labor by laying the responsibility for some of the flight southward
on the Taft-Hartley Act. At the same time, Kennedy tried to avoid directly
criticizing the South, though he did urge owners of the new southern-based
textile mills to boost wages. Realizing that New England was also a beneficiary of federal policies, he refrained from criticizing the South's inexpensive
hydroelectric power made possible under the TVA. But his assiduous efforts
to avoid offending southerners failed, causing him to have to clarify his views
in a subsequent article published in the *Atlantic* the following January. There
he freely acknowledged that "the modern plants and machines of the South,
and the new and vigorous ideas of southern manufacturers, set a standard
which New England industry should emulate, not try to destroy." With an
eye to his own political trajectory and the still-Democratic South, Kennedy
reassured southerners that he was more interested in helping New England
rehabilitate its own industry than in denigrating the South's many natural
advantages.[20]

Nor did Kennedy blame technology for the exodus of fifty thousand New
England textile workers. To the contrary, he claimed that those textile facilities that modernized their machinery, tried new production methods, and
adopted innovative ways to achieve greater efficiency were the ones that survived. Therefore, he argued that the federal government should incentivize
the purchase of new equipment to increase productivity and to allow for a
leveling of the regional playing field. While he exhibited faithfulness to New
Deal constituencies—urging an increase in unemployment compensation
and praising the central role that unions continued to play in Democratic
politics—Kennedy was not at all reluctant to embrace technology as critical
to raising productivity and lifting living standards. From these maiden floor
speeches forward, he would artfully straddle the divide between the benefits
of technology and its too-often-negative consequences for workers and their
communities.[21]

At the conclusion of his three days of floor speeches, Kennedy succinctly
summarized some thirty-five separate ideas for reinvigorating the New
England economy. He urged raising the minimum wage, creating a federal
reinsurance system to help supplement stressed state unemployment funds,
bolstering grants to small businesses, and accelerating amortization schedules
for the purchase of new modern equipment. But the centerpiece of Kennedy's program, and topping his list of recommendations, was his proposal for
the federal government to form regional development corporations (RDCs).

It was among the ideas that in modified form would later find its way into the Area Redevelopment Act, one of the first measures Kennedy would sign as president to address unacceptable levels of unemployment at the outset of his term. As he originally envisioned them, the RDCs were to be government-sponsored, businesslike operations designed to spur local economic development initiatives by partnering with local industry and labor. The RDCs would be empowered to provide technical assistance, issue favorable loans for business development, and engage in a range of consultative services, including tax strategies and analyses that could identify potential new business opportunities suited to individual regions. These regional centers would also assist failing businesses before it was too late. Indeed, Kennedy sought to take advantage of many local development initiatives that were already in existence around the country. He noted that unfortunately they were usually created after the departure of a major company or industry. Now, consistent with the liberal-labor coalition's emphasis on forecasts and anticipatory planning, Kennedy asserted that it was time to take preemptive steps to reduce the chances for future pockets of regional deterioration throughout the country.[22]

With Cold War adversary Russia now perceived as an ever-increasing technological threat, it was important for politicians with aspirations to higher office to demonstrate support for a strong national defense and, with it, the attendant backing of federal support for research into advanced new technologies. With his emphasis on the importance of government providing incentives for stressed manufacturers to modernize their equipment, Kennedy hinted at what became his later full embrace of technological advancement. Kennedy would campaign for the presidency criticizing the Eisenhower administration for the lackluster response to Sputnik and chiding Eisenhower for allowing the perceived missile gap with the Soviets. Later the Kennedy administration would galvanize support for a dedicated national space program and create a science office in the White House. With these maiden speeches on the Senate floor, Kennedy remained tethered to the core of the New Deal coalition's position on full employment. Yet he also signaled how far he was willing to embrace investment in new technology as central to regional economic recovery. Kennedy urged federal involvement in beefing up local economies by providing tax incentives for updating antiquated and uncompetitive industrial processes through new capital expenditures. And his redevelopment corporation model was one based on public-private cooperation rather than unconditional government largesse.

In these first senatorial speeches, Kennedy thus carefully balanced the importance of technology with the impact that modernization would have on jobs and communities, a vision he frequently articulated in his later quest for the presidency. As senator, though he helped recast the jobs-versus-technology debate and proposed creative new solutions, he did not lead the reform effort. Instead he joined others in the liberal-labor coalition, for example in endorsing Humphrey's effort to raise the minimum wage, and then in following Douglas's lead on broader depressed-areas legislation that incorporated many of Kennedy's suggestions, including the formation of the RDCs. In this postwar political milieu, however, such reform efforts were becoming more challenging to enact. After the minimum wage was finally raised in 1949, Humphrey would struggle to raise it again in 1955, and it would take Douglas three more Congresses before his legislation, inspired by Senator Kennedy, was finally signed into law by a delighted new President Kennedy. Throughout the fifties, though the liberal-labor coalition encountered formidable headwinds, it remained indefatigable in its quest to blunt the effects of technology on labor.[23]

Guaranteeing a Wage

Kennedy's singular focus on the attrition of jobs southward led him to propose remedies for damage already done. In contrast, Reuther sought to prevent workers in the vibrant auto industry from being subject to the emerging force of automation. Just as Wiener had urged him, and having achieved what he could at the bargaining table, he now urged Congress to act preemptively before it was too late. But Congress, like the public generally, remained largely uninformed about automation, and thus Reuther began his own education campaign, warning the political world that it needed to move ahead of the new phenomenon. For example, he used his testimony before the Joint Committee on the Economic Report in February of 1954 as his opportunity to sound the alarm. "You ought to go through the new Ford engine plant [the Brook Park Plant complex in Cleveland]. You can't find the workers," Reuther told the committee. He was convinced that politicians were not moving fast enough to understand how computers were dramatically transforming life on the factory floor. Throughout the midterm election year, Reuther continued his own campaign to raise the still ill-understood concept. He consistently urged policies that demonstrated the same imagination and creativity as displayed by the inventors of new technologies. The country remained in a brief recession and unemployment was hovering at 5 percent.

Indeed, the CIO claimed that over 1,750,000 workers were displaced from their jobs each year because of automation, and Reuther warned that with an estimated 750,000 new job entrants coming into the workforce annually, the unemployment situation was destined to become worse.[24]

Taking his own campaign before the CIO annual meeting in December 1954, Reuther declared that "social wisdom is called for to match scientific advances," and he unveiled his own idea for a greatly expanded guaranteed annual wage (GAW), going well beyond what the UAW had secured in the Treaty of Detroit. "We do not want to be paid for not working, but we do not want our people to be penalized when they haven't got a job through no fault of their own," Reuther told the rapt audience of some five thousand members. Unions needed to create an incentive for full employment, he said, by negotiating for companies to own up to their social obligations. His expanded GAW would guarantee worker salaries for an extended period once they were laid off. Reuther believed the new contractual obligation ensured not only fairness, but also that companies would consider the pace at which they automated.[25]

Though Reuther's aggressive advocacy for an expanded GAW landed him on the cover of *Time* magazine for a second time in June 1955, he was simply giving his full-throated endorsement to a concept necessary to bolster the existing federal unemployment compensation system. Despite the provisions in the Social Security Act that supported state unemployment compensation funds, federal support was still limited. In response, by the early fifties, a few companies were attaching a GAW-like feature to profit-sharing or wage-incentive programs, or even providing employee loans in down business periods. Even before Reuther struck his deal with the auto companies, the CIO had campaigned to make a GAW part of the 1946 full employment legislation. For its part, the Truman administration had recommended a three-month guaranteed wage. But absent legislation, the administration was then left trying to simply help the unions jawbone businesses to agree that a GAW should become an integral part of future labor contracts. The inhospitable politics quickly drove union leadership to make the GAW a central feature of labor contract negotiations. Then having achieved results at the bargaining table with the Treaty of Detroit, Reuther, though willing to continue to test the political waters, recognized that labor's best hope for an expanded GAW remained at the bargaining table.[26]

Thus, with UAW auto contracts up for renewal in 1955, Reuther confidently told assembled reporters at the National Press Club that achieving an expanded GAW was one of the union's priorities in the forthcoming negotiations with GM and Ford. But the resistance by industry was formidable. In the

postwar era, tough union-management negotiations assumed an almost ritu-
alistic aura. Threats of strikes became a part of the negotiating dance between
reluctant partners. Therefore, almost predictably, at the eleventh hour of a
threatened strike against Ford, Reuther proudly announced the union's victory
and gratuitously praised the company's foresight. In recognition of the realities of the automated workplace, the new contract established an expanded
$55 million GAW fund that would allow laid-off workers to draw a supplemen-
tal income benefit for up to twenty-six weeks. Ford also agreed to a five-cent-
per-hour contribution to an individual worker fund that could provide up to
twenty-five dollars a week for laid-off workers for half a year. When combined
with unemployment compensation, the amount represented about 65 percent
of an employee's regular pay, an important concession even as Reuther continued to argue for the union version of a more robust GAW. The precedent
set by the automakers quickly spread to the steelworkers. By August they
announced an agreement with the major can companies for a fifty-two-week
GAW. Eventually the UAW claimed that over a million of its workers benefited
from some form of guaranteed wage. Beyond the immediate relief it provided
to those out of work, it also acted as an incentive for companies to think twice
about how fast they introduced new machinery onto the factory floor. Since
a portion of the capital spent for new automatic equipment would now have
to be diverted to a fund that assisted unemployed workers, perhaps industry
would calibrate the pace of automation, as Reuther urged.[27]

The Automation Hearings, 1955

Fresh from this latest success negotiating an expanded GAW at the bargaining table, and now enjoying a growing national reputation, Reuther returned
to the political arena, stepping up his call to Congress to arrest the pace of
automation. The CIO's unparalleled grassroots effort in the 1954 off-year
elections catapulted the Democrats back to the majority in both the House
and Senate. Accordingly, organized labor did not waste any time pushing
its own agenda, including the growing threat from automation. Quick to
respond, and heralding its own intent to examine how automation was dramatically changing the economic landscape, the Joint Committee on the
Economic Report, now chaired by Wright Patman, author of the original
House version of the Full Employment Act legislation, reported that "there
is reason to believe the country is now faced with something in the nature of
an industrial revolution comparable to that introduced by the interchange-
able parts and the assembly line technique. We are told that literally millions
of both white collar and factory workers may be displaced in the matter of

a few years by the products of the electronic age." By October 1955, Patman announced that based on the wave of ominous news about automation, he would convene a special subcommittee to examine the topic of "automation and technical change." Concentrating on six specific industries, Patman focused particular attention on the competitive consequences of automation for his friends in small business, noting that with their limited resources, smaller enterprises would not be able to invest in the new automatic technologies. Throughout the hearings, the long-standing advocate for small business incessantly probed witnesses on the impact of automation and new technology on small enterprise.[28]

As the hearings proceeded, business and labor predictably staked out opposing positions. Reuther, with his usual visionary zeal, urged that government step up its forecasting capabilities to predict economic downturns so that measures could be taken preemptively to ameliorate the jobs impact. The liberal-labor coalition remained intent on finding a way to reinvigorate the idea of a more robust federal economic planning mechanism, like that jettisoned with the demise of the original full employment bill. Reuther repeatedly returned to this idea. He pointed to specific instances where technology was decimating workforces and communities. He also came with other recommendations, including reducing the workweek to thirty-five or even thirty hours, incentivizing workers to take early retirement, and increasing skills-based schooling for young employees and those about to enter the workforce. Yet while these measures might ameliorate the effects of automation, Reuther kept returning to the need to move ahead of the devastating impacts of the rapidly appearing new workplace technologies. He criticized Congress for being more intent on advancing automation by expanding the federal research budget than on tending to the needs of displaced workers. He believed the time had come to return to an honest discussion of how the federal government could better anticipate the consequences of accelerating automation.[29]

In stark contrast, the National Association of Manufacturers and its conservative allies continued to encourage a laissez-faire approach to technological development. In a refrain familiar since the Harding conference, its representatives reassured the subcommittee that any job losses would be offset by even more robust job creation. D. J. Davis, the vice president of manufacturing at Ford, defended automation, saying that it shifted the workforce toward higher-skilled jobs. He argued that the fears of automation were overblown. William Barton, the president of one of the companies producing the very automatic equipment at issue, declared that the elimination of human physical labor was both desirable and inevitable. He asserted that the trajectory of human progress was about to reach another milestone, with the new generation of machines finally eliminating man's subservience to machines altogether. The industry witnesses seemed to embody the techno-optimism reflected in a recent NAM pamphlet. Touting the benefits of an automated future, it concluded, "Guided by electronics, powered by atomic energy, geared to the smooth effortless workings of automation, the magic carpet ride of our free economy heads for distant and undreamed horizons."[30]

Witnesses from the Eisenhower administration shared the NAM's rosy view. They were bullish on how new scientific discovery and the technologies spawned by ever-expanding federal research budgets provided the very foundation for the ongoing economic expansion. Noting that one of the marvels of the postwar boom period was the growth in consumer spending, the administration's 1955 economic report enumerated the long list of consumer goods, such as modern homes, automobiles, radios, television sets, washing machines, air conditioning units, electric dryers, food freezers, and so on, that were dramatically improving the standard of living. Ironically, given Eisenhower's soon-jaundiced view of government's role in funding science, his 1955 economic report to Congress, issued five years before his far more critical parting address, acknowledged that "the Federal Government itself plans to spend more on scientific research and development during the coming fiscal year than it has done at any time in the past." Such an investment, the report optimistically concluded, would lead to even greater improvements in productivity and unparalleled prosperity.[31]

Not surprisingly, then, when Eisenhower's secretary of labor, James P. Mitchell, testified before the Patman subcommittee, he stated that automation posed little threat to jobs. Organized labor had long viewed Mitchell warmly. He opposed right-to-work laws, supported labor's right to organize, sympathized with the plight of migrant farmworkers, and advocated for better job opportunities for African Americans. Yet consistent with the administration's view, he saw automation as a force easily accommodated in the prosperous Eisenhower economy. It took time, he argued, to finance and bring online the new equipment necessary to automate. Thus, Mitchell tried to reassure skeptical committee members that during such a transitional period, a company could comfortably transfer some employees to other functions while retraining others. Accompanied by BLS chief Ewan Clague, the now-seasoned veteran of prior dialogues on technological unemployment, Mitchell painted the picture of a future where the number of boring, routine, and repetitive jobs would be eliminated. Recognizing that Patman supported the newly introduced House version of area redevelopment legislation (similar to Douglas's in the Senate), Mitchell adroitly directed the subcommittee's attention to what he characterized as isolated and localized pockets of unemployment. He claimed that these sprung up for idiosyncratic reasons having to do with noncompetitive aspects unique to local industries and therefore should not be attributed to automation. By restating the administration's binding faith in science and invention as the practical solution for the replacement of old and declining industries, the secretary suggested that these established industrial communities were capable of reinventing their economies. Moreover, he noted, the Department of Commerce had already established its own Office of Area Redevelopment to begin to address localized unemployment challenges. For the administration in this new era of prosperity, automation was viewed more as part of the solution than as a fundamental cause of unemployment.[32]

Perhaps the most surprising position offered at the extended hearings came from the one person most associated with popularizing the term "automation" in the first place. As a twenty-six-year-old Harvard business school graduate, John Diebold had adroitly summarized the new phenomenon in a book entitled *Automation*, published in 1952. By the time Patman convened his hearings, Diebold's thriving consulting firm at 430 Park Avenue boasted clients such as AT&T, Boeing, Xerox, and IBM. The acclaimed business consultant Peter Drucker wrote that Diebold defined automation in such a way that "recognized the arrival of not only a new technology, but also an altogether new way to look at work and the economy." Though he may have been expected to give his ringing endorsement to the views espoused by business and the administration, Diebold instead acknowledged the significant externalities associated with automation. Like Reuther, he urged Congress to think ahead beyond the immediate challenges of job training and relocating displaced employees. The corporate consultant wrestled with how the economy might absorb inevitable job losses and went so far as to acknowledge that automation would create problems of "greater difficulty than were overcome in achieving it." In his balanced testimony, the polished young Harvard graduate, who had been fired from his first corporate jobs for his too-futuristic predictions, tried to be clear eyed about automation's societal impacts. While he acknowledged to the subcommittee that automation presented challenges, he thought them to be manageable. He urged Congress to undertake a thorough study of its impacts and meticulously outlined what was required.[33]

Fittingly, on the final day of his hearings, Patman recognized a witness he described as "one of the grandfathers of the computers and numerical-control devices that we have heard about during the past two weeks." The month after Patman's hearing concluded, Vannevar Bush would announce his retirement from the Carnegie Foundation, from his board memberships, and from his government advisory roles. Thus, save for a few remarks at a spate of informal retirement dinners, his testimony to Patman's subcommittee represented one of his last official public statements. While Bush was generally optimistic regarding man's ability to control and accommodate automation, like Diebold, he offered his own cautionary note about that cohort of the population that would soon be pushed aside by the advance of these thinking machines. "Every time, in my opinion, that it [automation] pushes a man aside, because he is too old or too little endowed by nature to learn new skills, there is a social loss," Bush stated. As the inventor of one of the first analog computers, Bush applauded the committee for considering how many displaced workers would simply not be able to be retrained for the new automated economy. Such technological change, he warned, could halt careers in their tracks, compounding the stress on those workers still employed.[34]

Despite Diebold's and Bush's credible cautionary notes, however, there was little political momentum to arrest a phenomenon still difficult for most to fully grasp. Gross domestic product (GDP) was growing at a remarkable 7 percent, and an ebullient *Economic Report of the President* issued in January 1956 to Patman's joint committee summarized the previous year by highlighting the achievement of full employment, rising incomes, and a stable dollar. When the committee issued its own report distilling the findings of the subcommittee from the 644-page hearing transcript, it acknowledged that the economy was supporting full employment in spite of the steady advance of the automated workplace. Clearly there was no jobs crisis, and the joint committee saw no reason to recommend specific legislation to combat the impacts of automation. Consistent with the long held liberal-labor vision, however, the joint committee did assert that, "with careful planning" and with attention paid to the "timing of investment" in new technology, major layoffs and longer-term unemployment might be minimized.[35]

Depressed Areas Legislation

Despite the booming economy and rosy economic outlook, Senator Douglas and his liberal-labor allies continued pressing the case for their depressed areas legislation. As Secretary Mitchell had acknowledged, there were pockets of unacceptably high levels of regional unemployment. Eisenhower even called for a commission to address why unemployment remained high in certain regions. Much like the earlier Harding conference, the Eisenhower initiative, chaired by Inland Steel's CEO, was dominated by conservative business interests, and predictably concluded that "in a free economy, some displacement of workers and some injury to institutions is unavoidable." The administration utilized the commission's findings as its policy rationale for opposing trade adjustment assistance legislation offered by Senators Humphrey and Kennedy, and later as a way to resist Douglas's legislation. The administration's newly formed Area Development Division within the Department of Commerce provided conservative Republicans and southern Democrats with enough political cover to defeat Douglas's bill. Nonetheless, the Illinois senator would soldier on against seemingly impossible odds in an effort to move his own stand-alone depressed areas legislation.[36]

In January 1956, now wielding his own gavel as chairman of the Subcommittee on Labor, Douglas held a relentless eighteen days of hearings. As part of the deliberations, the Illinois senator took the subcommittee to small towns in his home state. From his days as a professor of economics to his role as a New Deal adviser, Douglas had seldom put much emphasis on the role technology played in causing unemployment. Nor, apparently, was he prepared to do so in his new capacity as a subcommittee chairman. Indeed, in some 1,300 pages of testimony, the term "automation," still making its way into the American vernacular, was invoked only sparingly. There were notable exceptions from familiar quarters. The Democratic governor of Michigan, Mennen "Soapy" Williams, noted that although automation was merely a word to many in the state, it was already taking its toll. The charismatic heir to a cosmetic company fortune (thus his nickname) predicted that computer technology would require a majority of workers to learn new skills. The governor forecast that within a generation, four out of five workers would be displaced from their jobs because of automation. Two other Democrats, West Virginia senator Matthew M. Neely and Pennsylvania governor George M. Leader, witnessing the impacts of automation on their home-state coal and steel industries, echoed Mennen's concerns. Yet as the hearings proceeded, such attention to automation was the exception rather than the rule. Douglas himself never mentioned the term. Senator Humphrey, always looking for the silver lining, acknowledged that even though automation was a new concept, he was reassured by Patman's hearings on automation the year before that it would lead to the creation of jobs as well—though he quickly added the caveat that he was less than confident there would be an overall net jobs gain. For his part, Douglas seemed intent on focusing attention on the continuing unacceptable levels of employment in certain areas, rather than dwell on the causes. And he focused attention on the need for remedies rather than on problem analysis.[37]

The companion to Douglas's bill was sponsored in the House by Congressman Daniel Flood (D-PA), who represented a coal-producing area of northeastern Pennsylvania increasingly hollowed out by the automation of the mines. Flood was a bona fide congressional character. Once a Shakespearian actor, he sported a waxed mustache and impeccable dress, landing him the nickname "Dapper Dan." Particularly close to Senator Kennedy, Flood had mentored him as a young Massachusetts congressman. With House offices next to one another, the two would remain close friends even after Kennedy moved to the Senate. Following the template outlined in Kennedy's maiden speeches, the Douglas-Flood Area Redevelopment legislation addressed the immediate employee fallout from declining industries, while seeking to spur declining businesses to modernize. The proposed legislation funded state manpower training programs and bolstered state worker compensation systems. Not only did it contain the financial wherewithal to help qualifying communities to construct new industrial facilities, but it also provided favorable tax treatment for the modernization of uncompetitive factories. As Flood's Pennsylvania colleague and former AFL representative George Rhodes (D-PA) warned his colleagues, "Either we adopt public policies to study, anticipate, and channel the course of automation toward human betterment, or like the Frankenstein monster, it may rise up to destroy us."[38]

As they soldiered forward against the odds in their efforts to enact legislation in the 85th Congress (1957–58), Douglas, Flood, and their liberal-labor allies in both chambers received a jolt of political momentum in the form of what became known as the Eisenhower recession. Though it began in August 1957 and then quickly ended in April 1958, the economic downturn lingered on in many areas of the country. Unemployment rates in many depressed regions grew worse. In the coal town of Beckley, West Virginia, by August 1958 unemployment stood at 25 percent. In the steel town of Wilkes-Barre, Pennsylvania, it topped 18 percent. In the nation's automobile capital of Detroit, it was over 14 percent. There were an estimated one hundred similarly depressed communities nationwide. During a hearing on Flood's bill, Governor Leader of Pennsylvania noted that the state's heavily automated steel industry was now producing one-third more steel than it did after World War II and doing so with far fewer employees. Kentucky governor Ben "Happy" Chandler, eventually best known as the baseball commissioner who allowed for the sport's integration, told Douglas's subcommittee that since 1950 the number of coal miners in Kentucky was reduced from forty-eight thousand to twenty-nine thousand as the result of automation.[39]

In the suddenly more favorable political context during the spring and summer of 1958, Douglas and Flood, sensing the opportunity to compromise with the administration, agreed to drop the idea of federal grants, instead establishing a revolving loan fund for depressed communities. The Senate narrowly passed their modified bill in May by ten votes, and then just days prior to adjournment, it squeaked by in the House with eighteen votes. The administration, however, took note of the force of the opposition in both chambers, and while members were back home campaigning, the president quietly pocket vetoed the measure. Eisenhower made the case that his own administration's redevelopment office was adequately supporting state and local management of these regional challenges, and thus there was no need for additional bureaucracy. The federal government's role, the administration argued, should be supportive, not central.[40]

A Special Committee

Yet the administration seemed to have miscalculated the extent of public concern. With so many communities across the country impacted by lingering postrecession unemployment, Eisenhower's pocket veto was considered by some to be a contributing factor in the Democrats' midterm election triumph. The Democrats widened their majority margins substantially in both chambers, going from a five- to a ten-seat margin in the Senate, and from a thirty-nine- to a sixty-nine-seat advantage in the House. Yet as the 86th Congress began, Douglas and Flood realized that satisfying the stubborn White House remained their greatest political obstacle. Seeking to put additional pressure on the administration for Douglas's bill, Johnson, now the majority leader, named a new freshman senator from Minnesota, Eugene McCarthy, as chairman of a special committee formed to address unemployment. Johnson then stacked the committee with northern-state liberal-labor Democrats. If nothing else, the new committee would be a platform for the party's liberal wing to continue to highlight the ills stemming from automation and to continue to pressure the lame-duck Eisenhower administration. Beginning in October 1959, McCarthy, who, like Douglas, had also once been an economics professor, welcomed the first of 538 witnesses who would testify over twenty-seven days of hearings on unemployment held across the country.[41]

McCarthy opened the hearings by noting that a new form of unemployment, so-called "structural unemployment," could be attributed to the ever-advancing new automated technology that now seemed baked into the economy. He observed how, despite unparalleled prosperity since the end of the Korean War, unemployment continued to be above the designated 4 percent level generally accepted as representing full employment. The committee's deliberations would publicly highlight how, in many depressed areas of the country, joblessness remained unacceptably high. The committee traveled to the coal regions near Rock Springs, Wyoming, to hear about how automation was changing mining practices. It was a theme repeated in the industrial Midwest, where auto parts suppliers in South Bend described how they were going out of business because they could not compete with automated shops. The special committee also visited timber country in the Northwest, where new machinery was putting loggers out of work.[42]

Perhaps nowhere was the impact of automation on jobs more deeply felt than in West Virginia, where committee member Senator Jennings Randolph (D-WV) took the lead in sympathizing with his beleaguered constituents. Beginning in 1937, the state's coal industry produced some 455 million tons of coal and employed 491,864 employees. In 1958 it produced 410 million tons yet needed only 195,000 employees. The technological acceleration in coal extraction resulted in an 80 percent rise in average productivity. When the special committee arrived in Beckley on November 16, 1959, it was in the center of Congressman John Slack's district. While the rest of the nation recovering from the recession was reveling in a GNP growing at an annual rate of 6.6 percent, in Slack's once-prosperous district, well-paying coal jobs were being eliminated in staggering numbers by new automatic machinery. As Slack stated plainly for his colleagues, these jobs were simply gone. "I offer for your approval a new word to be added to the English language," he said. "That word is 'gone employment.'"[43]

McCarthy's committee would issue its findings six weeks before what turned out to be West Virginia's pivotal May 10, 1960, Democratic presidential primary. The report concluded that the baby boomer generation coming into the workforce in the next decade would further exacerbate the already significant impact of new technology on jobs that fell hardest on the young and African Americans. According to McCarthy's committee, unemployment was now a class issue, not a mass issue. The effects fell hardest on older workers, nonwhites, women, the unskilled, and the least educated. "If a city the size of Detroit can be shaken by automation, then no community in the country can be complacent about its effects," the report warned. The committee expressed unequivocal concern that automation was a looming force complicating solutions to unemployment, whatever the other underlying causes. Unlike more discernible factors, such as trade or the cancellation of government contracts, automation seemed a more insidious force that served to exacerbate the jobless situation. More often than not, it was becoming the primary cause. Though exhaustive in its review and acknowledging automation as a principal cause of joblessness, however,

McCarthy's panel seemed unable to recommend any innovative new policies that would address technology's advancement. It offered the central economic budgeting and planning idea previously jettisoned during the debate over full employment legislation, and then defaulted to a series of recommendations that mirrored New Deal policies to stimulate job creation, such as investment in public works and funding of emergency public housing.[44]

In contrast to the majority's ideas, a group of Republicans who served on McCarthy's committee issued their own recommendations. With liberal Republicans such as Winston Prouty (R-VT) and John Cooper (R-KY) taking the lead, they recommended measures to protect minorities by expanding the criteria for eligibility for unemployment compensation. They further suggested that states expand the tax base from which they could levy the taxes that supported their unemployment funds. Their minority report also called for raising the minimum wage, though not to the level the Democrats proposed, and it urged the administration to bolster its aid to depressed areas. In sum, it was indicative of how the moderate and liberal wings of the Republican Party often served as a foil to more conservative southern Democrats and the new wave of conservative Sun Belt Republicans. At the conclusion of their report, these progressive Republicans further recommended that a national commission be convened on automation. They suggested that if the Democratic majority could not develop anything more imaginative to address the automated future, perhaps a blue-ribbon panel, insulated from political pressure, might do better. It seemed a tacit admission that as politics kept chasing after the effect of new workplace technology, it was simply running out of new ideas.[45]

Before the depressed areas legislation made its way to the White House again in May 1960, Douglas made several additional attempts to modify his bill so that it might overcome Eisenhower's objections. Yet even this substantially modified version would be vetoed by the president. But because Congress was in session, majority leader Johnson immediately scheduled a vote to override. Despite support from Republican senators Cooper and Scott, the override fell eleven votes short. Southern Democrats, including John McClellan, chairman of the Labor Committee, sided with Eisenhower, reflecting the continuing challenge faced by the liberal-labor coalition even when there were Democratic majorities in both chambers. McClellan and his colleagues in the liberal-labor coalition found the political climate increasingly inhospitable to policies that addressed the most egregious examples of how the new automated technology was taking jobs. Frustrated in his repeated attempts to negotiate with the Eisenhower White House over three successive Congresses, Douglas was more determined than ever to help his colleague John Kennedy win his quest for the presidency. Secure in his own reelection in 1960, Douglas logged forty thousand miles, delivered almost three hundred speeches, and reached an audience estimated at some half a million to help turn Illinois for Kennedy. As the campaign for president unfurled in the spring of 1960, Kennedy, moved by the unemployment he had witnessed in West Virginia, joined his friend Senator Cooper and those other progressive Republicans on McCarthy's committee by suggesting that perhaps it was time for a presidential commission to think ahead about the economic consequences of automation—something the politics of the moment seemed incapable of doing.[46]

FIGURE 1. Walter Reuther addresses an audience at the United Auto Workers West Side Local 174, Detroit, Michigan (circa late 1930s). Courtesy Walter P. Reuther Library, Wayne State University.

FIGURE 2. Frances Perkins signing a document in 1918. Courtesy Library of Congress.

FIGURE 3. President Harry Truman signs the Employment Act of 1946 on February 20, 1946. Present are (left to right) Senator Joseph C. O'Mahoney (D-WY), Rep. Wright Patman (D-TX), Sen. William Langer (R-ND), Rep. George Outland (D-CA), Rep. William M. Whittington (D-MS), Sen. George D. Aiken (R-VT), Rep. Carter Manasco (D-AL), Sen. George I. Ratcliffe (D-MD), Rep. George H. Bender (R-OH), Sen. Abe Murdock (D-UT). Courtesy Associated Press; AP Photo by William J. Smith.

FIGURE 4. President Franklin D. Roosevelt signs the Social Security Act on August 14, 1935. Present are (left to right) Rep. Robert Doughton (D-NC), Sen. Robert Wagner (D-NY), Rep. John Dingell (D-MI), Rep. Joshua Twing Brooks (D-PA), Secretary of Labor Frances Perkins, Sen. Pat Harrison (D-MS), and Rep. David Lewis (D-MD). Courtesy FDR Library.

FIGURE 5. President John F. Kennedy signs the Area Redevelopment Act of 1961 in the Oval Office on May 1, 1961. Present are (left to right) Rep. Daniel J. Flood (D-PA); Rep. Cleveland M. Bailey (D-WV); Sen. Hubert Humphrey (D-MN); Vice President Lyndon B. Johnson; Rep. Frank Stubblefield (D-KY); Sen. Paul H. Douglas (D-IL); Rep. Frank Chelf (D-KY), behind Douglas' shoulder; Sen. John Sherman Cooper (R-KY), in back; Rep. Brent Spence (D-KY); Sen. Robert C. Byrd (D-WV); Sen. Joseph S. Clark (D-PA); unidentified man; Rep. Carl D. Perkins (D-KY). Courtesy JFK Library.

FIGURE 6. Vannevar Bush standing before his elaborate "differential analyzer," which debuted in 1931. Courtesy Computer History Museum.

FIGURE 7. Isador Lubin (left) briefing Senator Joseph C. O'Mahoney (D-WY) prior to a hearing of the Temporary National Economic Committee, formed by Congress in 1938 to examine monopolistic behavior and decipher the causes of the Depression, including technological unemployment. Courtesy Library of Congress.

FIGURE 8. Howard Bowen, chairman of the Automation Commission. Courtesy Frederick W. Kent Collection, University of Iowa Libraries.

FIGURE 9. Anna Rosenberg. Courtesy Library of Congress

FIGURE 10. President Lyndon Johnson observes a student at the Camp Gary Job Corps Center in San Marcos, Texas, on November 8, 1965. Courtesy LBJ Library; photo by Frank Wolfe.

FIGURE 11. Norbert Wiener. Courtesy Estate of Francis Bello/Science Photo Library.

CHAPTER 6

Creating a Commission

It is my intention to appoint a Presidential Commission on Automation, composed of the ablest men in public and private life.

—John F. Kennedy, remarks to Congress, July 22, 1963

Shortly before 12:30 p.m. on January 2, 1960, Senator Kennedy left his office in room 362 of the Old Senate Office Building and walked down the marble corridor to the caucus room, overflowing with excited supporters and an eager press corps gathered to hear him declare that he was a candidate for president of the United States. His succinct announcement was neatly typed on a single sheet of paper. In his trademark cadence, he spoke about bolstering science to ensure the nation's technological edge. Then he addressed the scourge of unemployment across the country, from the farmland to the cities. In one brief statement, he spoke to two core American values that continued to compete for political attention. The challenge, he acknowledged, was how to expand the economy for the benefit of all Americans. The brief announcement took less than five minutes to deliver, but the senator was off and running.[1]

On March 8, Kennedy convincingly won the New Hampshire primary. A month later, he pulled off a stunning upset in the Wisconsin primary, defeating the neighboring state's senator, his friend and colleague Hubert Humphrey. The nation's political attention then turned to the next primary battleground, West Virginia. During a grueling campaign, Kennedy found himself visiting the very coal mines that were the focus of Senator Randolph's hearings on unemployment there the year before. As the primary battle intensified, Kennedy's Catholicism emerged as an issue and his poll numbers plummeted. The preprimary election polls grew closer by the day. West Virginia's primary suddenly riveted the nation's attention and consumed all of Kennedy's energy.[2]

In a feature piece for the *New York Times*, its acclaimed reporter Harrison Salisbury wrote that the battle would be won by the candidate who best connected with the state's coal workers, who were reeling from automation. Salisbury reported that it took only 40,000 workers in 1960 to extract approximately the same amount of coal that required 117,000 miners in 1950. As Kennedy toured the state's struggling mining towns, he came to realize the human toll exacted by the new coal extraction technologies. He toured places like Mullins that had never seen a presidential candidate. He went to Logan, where twenty years before, 15,000 men had worked in the mines. Now there were only 5,500. Visiting a mine in Stanford just as the day shift ended, he greeted some sixty miners as they made their way out of the shaft. "You John Kennedy?" one of the blackened miners said.

"Yes," said the senator in his neat, gray pinstripe suit.

"Okay, I'm your man," responded the miner, and the senator flashed his easy smile. These scenes repeated themselves in the final days of what turned out to be Kennedy's decisive 22 percent primary victory over Humphrey.[3]

By the time he arrived in Grand Rapids on June 6 to address the AFL-CIO's summer meeting, Kennedy was more confident than ever that he would win the Democratic Party's nomination for president. Just a few days before, he had met with Michigan governor Mennen Williams at the Grand Hotel on Mackinac Island and secured his enthusiastic endorsement. The announcement was a critical piece of Kennedy's closing strategy. Despite the senator's outward confidence and his previous primary victories, the nomination at the Democratic convention was a month away and the outcome remained very much in doubt. At least Kennedy could now rest assured that the governor of the nation's most unionized state would be arriving in Los Angeles on July 11 with most of Michigan's 51 delegates securely in the Kennedy column. With these delegates in hand, Kennedy operatives counted over 600 delegates pledged or bound to support him. He needed a total of 761 to win the nomination.[4]

Kennedy was determined to win the endorsement of organized labor. His Grand Rapids AFL-CIO speech thus became yet another pivotal moment in the campaign. Fortunately for the Kennedy camp, though Reuther remained officially undeclared until the nomination was secured, he was privately urging Kennedy's nomination. In fact, it was speculated that Reuther was behind Governor Williams's endorsement. Not surprisingly, then, in preparation for his appearance at the AFL-CIO meeting, Kennedy's team mapped out a speech on the challenges of automation that utilized Reuther's positions as essential points. Since he first raised automation in his testimony before the Joint Committee on the Economic Report in 1954, the UAW president had avoided addressing whether automation meant more or fewer jobs in the aggregate. With so many previous analyses unable to answer that question, Reuther instead used specific examples of jobs lost and the immediate impact it had on families and communities. In fact, his approach was not dissimilar to Kennedy's in describing the economically depressed New England textile industry in 1953. Reuther wanted workers to realize their share of the benefits derived from increased productivity resulting from automation, and he therefore continued to vigorously promote an expanded GAW. As the decade turned, Reuther also began to focus on how machines changed control of the factory floor, including how work assignments were designated, the pace and nature of process changes, and the very substance of production standards themselves. And he returned to his familiar theme of urging government to devise a way to better forecast technological advancement and then plan for the consequences of growing worker displacement in industries that were targets for automation. In short, he called on the government to accelerate the development of forward-looking policies that captured the reach of technological innovation.[5]

Well briefed on Reuther's extensive position on automation, Kennedy received a rousing, foot-stomping welcome from some 1,300 union members as he entered the Grand Rapids Civic Auditorium. When the speech ended, Kennedy was mobbed by the delegates and would need local police to escort him to a separate room for yet another press conference. The energetic young candidate repeatedly brought the enthusiastic union members to their feet with his soaring rhetoric: "Today we stand on the threshold of a new industrial revolution—the revolution of automation. This is a revolution bright with the hope of a new prosperity for labor and a new abundance for America—but it is also a revolution which carries the dark menace of industrial dislocation, increasing unemployment, and deepening poverty." Kennedy was both focused and forceful in expressing his concerns over the impacts of automation on the blue-collar worker, but he also struck a balance, mindful that white-collar business leaders would also read his speech with interest. He described what the audience instinctively knew, that as "modern technology continues its inevitable advance, thousands of processes and functions now performed by men, will be done, more cheaply and more efficiently by machines."[6]

To manage the new automated reality, Kennedy recommended that for its part, management face up to the impacts on workers. Modulating the pace of automation needed to become an integral part of the collective bargaining process. Moreover, management needed to help workers displaced by automation to find new work. Kennedy urged that companies step up their retraining efforts for the unemployed and increase unemployment compensation throughout the transition to new jobs. He further suggested that it was time to convene a national conference of business and labor leaders to "seek solutions to the problems of automation and reaffirm the determination of labor and management to make full use of modern technology in order to protect the interests of affected employees while the conversion is being made." Like Reuther, Kennedy advocated that the benefits of technology be shared, and he thought a congressionally authorized presidential commission might provide a roadmap for how to achieve a more equitable distribution.[7]

Repeatedly on the campaign trail, Kennedy spoke to what he called "the new industrial frontier." When he arrived at a Weyerhaeuser plant in Eugene, Oregon, he urged government, industry, and labor to work together to make automation work for everyone. "Workers," he said, "must be retrained— factories must be relocated—and above all we must be certain that the worker himself shares in the fruits of his increased productivity." In Pocatello, Idaho, he rallied the blue-collar crowd by telling them that he saw among them no sense of defeat over automation, but rather "a determined spirit that has made this nation the greatest nation on earth." In Milwaukee he borrowed directly from Reuther's belief that automation allowed for workers to realize greater economic gain, particularly if the government interjected itself at the front end with greater planning. Kennedy observed that throughout the fifties, as automation ramped up throughout industry, there had been "no planning for automation." As liberal-labor reformers had long insisted, the pace of technological change called for a government entity that might forecast the potential impacts on jobs and communities from these emerging new technologies. Rekindling the fundamental liberal-labor aspiration, Kennedy promised that would change if he were elected. "This is not the same nation it was in 1952," he observed. "Automation was little more than a word then—now it means unemployment and hardship for untold thousands of workers." During a later debate, when Republican candidate Richard Nixon tried to suggest that the unemployment issue would not be significant until the number of jobless exceeded 4.5 million, Kennedy, flashing his trademark wit, responded that such a rationale might not be convincing to the 4,499,999 who were currently unemployed.[8]

New Frontiers

Even the outgoing president was becoming more concerned about the development of a technological culture, dominated by military interests, that he came to believe was having a distortive impact and creating an autonomous "military-industrial complex." Just three days before Kennedy's January 20, 1961, inauguration, President Eisenhower bid farewell to the nation with a prescient warning about the effects of this phenomenon on the nation's research community. Having presided over a steady growth in federal research funding, much of it directed at Cold War defense needs, Eisenhower now warned of much the same danger that Bush had forecast when he wrote what became known as his "Endless Frontier" memo to FDR.[9]

In contrast, the incoming president quickly pivoted to the brighter side of scientific promise. Beginning with his inaugural address, Kennedy sought to redefine the United States' scientific advantage in the ongoing Cold War. From landing a man on the moon within the span of a decade to spreading the nation's technological abundance abroad through the Peace Corps, the energetic and idealistic administration sought to reaffirm the promise of US technology. Kennedy saw the nation's vast technological capability as critical to helping build the economies of developing nations and thus as a promising means to winning the Cold War. Yet, continuing to balance the competing principles of the nation's political economy, his very embrace of technology did not dampen his continuing concerns over the impacts of automation on US laborers.[10]

As Kennedy assumed office, the country was pulling out of yet another brief recession, which had begun in April 1960. It was mild enough that the outgoing administration chose to ignore it altogether. Eisenhower's final economic report perfunctorily noted that in the second half of the year, production declined and unemployment rose. Indeed, the economic signs were mixed, and the *Wall Street Journal* called the downturn the "Peculiar Recession." The *Journal* reported that, indeed, not all sectors of the economy were experiencing a downturn. In fact, production in many sectors was up. Coming on the heels of a deeper recession in 1958, however, others speculated that perhaps if economic conditions did not constitute a full-blown recession, they at least represented a stagnation that had finally set in after the extended postwar boom. From a practical standpoint, what puzzled Kennedy and his economic advisers was that the increase in output that began in the second quarter of 1961 was not being accompanied by any commensurate gains in employment. After the election, unemployment rose. By the end of Kennedy's first year as president, it stood as high as 6.7 percent, well above the desired 4 percent. Though the nation's GNP would grow at a rate nearing 5 percent throughout Kennedy's term, concerns over unemployment shadowed his presidency.[11]

In Kennedy's one thousand days in office, the employment issue would remain a stubborn riddle and reinforce his commitment to form a commission in order that the country's best and brightest could examine automation from top to bottom. At the center of the administration's effort to thwart the impacts of automation stood Kennedy's secretary of labor, Arthur Goldberg. Goldberg saw the nation in the throes of a second Industrial Revolution that necessitated that it think anew. "To define the problem is also to state the promise, and to envision the progress is to define the problem," he observed as he proceeded to advocate a series of administration initiatives. Since the neutering of full employment legislation, the liberal-labor coalition had to be content with eking out incremental improvements to the minimum wage and unemployment compensation. Now Kennedy found in his secretary of labor someone who reinvigorated the liberal-labor coalition's workers agenda and recast the debate between technology and jobs.[12]

Shortly after the inauguration, Goldberg embarked on a five-state tour of the Midwest to examine the unemployment situation. At the end of his twenty-five-stop whirlwind, Goldberg estimated that soon the unemployment rate would be the highest since the Depression. During the highly publicized tour, the secretary built the case for a second incarnation of a Full Employment Act. When he visited Detroit, Reuther told him that eighty thousand jobs would have to be provided every week simply to offset the loss of jobs occurring because of automation. Whether this was an accurate projection or not, Goldberg was concerned about a looming jobs crisis and called for the quick enactment of the president's plan to extend unemployment compensation benefits. In Columbus, Ohio, Goldberg confronted business leaders who warned that the proposed federal supplement to unemployment compensation would result in added payroll taxes. Goldberg wagged his finger at the head of the Ohio Manufacturers Association, noting that he had yet to find an affluent unemployed person. In fact, Goldberg told the less-than-sympathetic audience, the administration's proposal did not go far enough, and he further announced that there would be no compromising. His own office had drafted a new full employment bill that it hoped to convince the administration to pursue. Meanwhile, the president signed a stopgap measure to extend federal aid to those states where unemployment compensation funds were already beginning to run low.[13]

As Goldberg publicly pressed the administration's case, the new president quietly helped his old friend Senator Douglas to at last enact the long-pending depressed areas legislation. After failing in three previous Congresses and facing still-narrow Democratic majorities in the 87th Congress, enactment would again require broad bipartisan support to overcome the predictable resistance of some southern Democrats. The area redevelopment bill that Kennedy would finally sign genuflected to their concerns by locating the Area Redevelopment Agency in the Department of Commerce, where Eisenhower's earlier initiative resided. It also subjected the new office to annual appropriations rather than unlimited Treasury funding. Additionally, to placate many southern congressmen, it not only equalized the number of rural areas targeted for assistance but also required that federal money to states and localities come in the form of loans rather than outright grants. Though he had compromised to adjust to the new political context, Douglas stood proudly at Kennedy's side, along with the president's old friend and lead House sponsor Congressman Flood, at the May 1, 1961, Oval Office signing ceremony. In its account, the *New York Times* tied Kennedy's support directly back to campaign promises he had made in West Virginia the year before. "In this free society, we want to make it possible for everyone to find a job who wants to work," he pledged. Since it was Kennedy himself who had first raised the issue on the Senate floor in 1953, it seemed fitting that Douglas's depressed areas bill was one of the first major bills the new president signed.[14]

By August the liberal-labor coalition claimed another victory, with the Senate passage of the Manpower Development and Training Act (MDTA). Introduced by Senator Joseph Clark (D-PA), the bill addressed the long-recognized need to establish a federally supported program of manpower training to bolster uneven state training initiatives. Lyndon Johnson's earlier commitment to forming the Special Committee on Unemployment Problems was credited with developing the political impetus and substantive record necessary to help push Clark's measure forward. Though the manpower bill would briefly stall in the Rules Committee, by March 15, 1962, it would pass the full House and was on its way to the president, who had long advocated for just such a program. In signing the bill, Kennedy noted that by ensuring fifty-two weeks of retraining for displaced employees, the new law helped to restore the commitments to job retraining made during the debate over the Employment Act of 1946. In addition to unemployment compensation for displaced workers, those laid off could now receive supplemental allowances while they retrained for new jobs. Predictably, much of the debate surrounding Clark's legislation swirled around the manpower crisis caused by automation, and the need to retrain workers. Automation's actual net impacts on jobs remained as murky as the effects of technological unemployment had been to earlier policymakers. But the MDTA required the secretary of labor to evaluate the overall jobs impacts and related loss of worker benefits and societal problems caused by automation. To assist with such a tall task, the bill established an advisory council that Goldberg hoped would not only evaluate the current impacts of automation but also provide recommendations that would allow the department to finally get ahead of the technology curve. In his signing statement, the President, taking some license regarding the actual scope of the legislation, suggested that "this far-reaching bill not only addresses itself to the problems of the present, but requires us to anticipate future needs as employment conditions change." While nothing in the bill's language specifically directed such forecasting, the president was mindful of his secretary's doggedness in pursuit of the liberal-labor coalitions long-sought yet still-unfulfilled goal.[15]

Indeed, prior to taking a seat on the Supreme Court, Goldberg was a relentless advocate inside the administration for finding any way possible to thwart automation's impacts. Soon the new administration was doing just the sort of analysis Goldberg advised. By January 1962, Kennedy's Advisory Committee on Labor-Management Policy issued its first assessment, titled *The Benefits and Problems Incident to Automation and Other Technological Advances*. The committee, which Goldberg cochaired with Commerce Secretary Luther Hodges, included a dizzying who's who of US labor and business leaders. In their report, the members struck what was becoming the usual equipoise between the virtues of automation in spurring economic growth and its deleterious impacts on jobs and communities. Though automation was presented as an essential component of US economic growth and a linchpin of the claim of technological superiority during the Cold War, the report acknowledged that it could not be allowed to simply run its course. The panel observed that even with the improved productivity that resulted from automated operations, the nation's economic growth was insufficient to support the job growth necessary to overcome automation's overall impacts on jobs. Moreover, the group expressed concern that public employment opportunities were insufficient to the task of relocating displaced employees. Further complicating the situation, displaced employees often did not want to relocate, and the performance of educational and manpower training capabilities remained uneven across regions, a reality the new MDTA hopefully would rectify.[16]

While the panel's analysis seemed to reflect current conditions, many of its recommendations consisted of reinventing New Deal programs such as public works projects or a shortened workweek. In addition to obligatory nods to expanding unemployment compensation and raising the minimum wage, the administration report argued for fiscal policies that would spur higher economic growth rates—namely, the proposed Kennedy tax cuts. Meanwhile, management was called on to give employees more adequate notice before layoffs, to protect pensions and other benefits during transition periods, and to better coordinate with union and governmental authorities in relocation and retraining efforts. Businesses were also asked to provide the workforce with greater flexibility (e.g., early retirement, relocation to new jobs, etc.), compensation during job transitioning periods, and greater worker mobility, and to end discriminatory hiring and retention policies. For their part, unions were asked to eliminate discrimination against members because of sex, race, or age. And as with analysis dating back to Hoover's Presidential Advisory Committee on Unemployment, the panel called for an analysis that might help disentangle the impact of automation from other factors that impacted unemployment. As for sorting out more precisely the impact that automation was having on jobs, the report remained stumped when it came to rendering any precise determination of technology's role in overall unemployment. If there were some improved measures of the rate at which new technologies were being adopted in specific industries, then perhaps policymakers might get ahead of the unemployment curve. This was precisely the sort of information that the proponents of the original Full Employment Act, with their National Production and Employment Rights Budget, sought yet failed to institutionalize.[17]

Perhaps frustrated by the seemingly conventional recommendations that the advisory panel made, Goldberg turned to the director of the Bureau of Labor Statistics, Ewan Clague, a former adviser to Senator Wagner, to begin to identify what new innovative processes, controls, and machines were in their early stages of development so that the administration might gauge their potential impact on employment. For the nonpartisan, professional Clague, who had survived several presidential transitions, it must have seemed like déjà vu all over again. Yet, like the BLS's previous efforts along these lines, its report entitled *Technological Trends in 36 Major American Industries* failed to reach any overarching conclusions. It did explain how new technologies were changing operations in specific industrial sectors, and it also rendered some broad generalizations about what automation might mean for labor requirements in the near term. Yet, again, the BLS analysis proved just a snapshot in time, even as Goldberg believed that, armed with such

information, the federal government might design policies that could alleviate the worst effects of the sort of structural unemployment that increasingly seemed baked into the economy.[18]

For example, the BLS noted that in the very same textile industry that
had so concerned then-freshman Senator Kennedy in 1953, accelerated
depreciation schedules for capital expenditure had accelerated the adoption
of electronically controlled high-speed machines and automatic looms with
continuous automated spinning. The wave of capital spending was further
reducing industry's labor needs. In addition, the emergence of international
competition made further modernization of the US industry inevitable, even
if it also meant job loss. Estimates as to how much, when, and where it
would occur remained as much a matter of speculation as when Kennedy
first registered his concerns in 1953. Similarly, in the coal-producing regions
that occasioned Senators Randolph and Clark's concerns during the special
committee's field hearings in 1959, the report predicted a continued decline
in employment because of mechanization. Analysts forecast that a shift from
conventional "room and pillar" mining would in the coming decade give way
to long-wall mining and thus further deplete worker ranks—the very "gone
jobs" that Congressman Slack had referred to in 1959. The report indicated
that by the end of the decade, almost half of production operations would
no longer require workers. In the timber industry, which then-candidate
Kennedy had visited a few months after McCarthy's field hearings in Eugene,
Oregon, the BLS reported that new machines such as mechanical lumber-
jacks were transforming the way in which trees were sawed, logs stacked,
and lumber cut. In all of the thirty-six industries under the special committee's review, employment was predicted to continue to decline, but precisely
by how much and when, and what could be done to prevent it, all remained
too much for analysis or the political imagination.[19]

The Indefatigable

While the administration attempted to develop its own new approaches
that might get in front of automation, Congress also opened new inqui-
ries into policies that might help alleviate automation's effects. On May 20,
1963, fresh from his victory on the MDTA and now exercising the oversight
authority of the act, Senator Clark gaveled in the first of forty-nine days
of hearings entitled "Nation's Manpower Revolution." Clark's Senate Subcommittee on Labor and Public Welfare would amass a 3,347-page hearing
record attempting to further plumb the depths of the unemployment problem. Clark acknowledged that despite the best efforts of both Congress and

the administration, the nation confronted the paradox of having four million unemployed, while also suffering severe shortages of skilled workers to operate highly complex technologies. In his opening statement, Clark noted that economic growth alone would not ameliorate new "structural unemployment," the term increasingly used to describe the so-called long-term unemployment that resulted from automated assembly lines. In the new automated economy, frictional unemployment (a term typically associated with workers transitioning to new job opportunities) had suddenly assumed a disturbing new dimension. Automation was creating a burgeoning class of those pushed aside by machine technology and in need of work. Not only was retraining displaced employees to learn new skills imperative, but so, too, was relocating the highly skilled to fill new jobs in emerging job-rich areas.[20]

The parade of administration spokespersons appearing before Clark's subcommittee seized the opportunity to emphasize how the tax cuts the administration now aggressively sought might help alleviate the stubbornly high unemployment numbers. Punctuating the administration's position, Walter Heller, chairman of the president's National Council of Economic Advisers, testified that even with the recently enacted MDTA, it took time for displaced workers to learn the new skills they needed and to identify new work opportunities. Indeed, the MDTA provisions provided the out of work with up to a year of retraining funds. Meanwhile, Heller noted, many communities were suffering from an ill-equipped employee base. Heller was one of those scholars who believed there was a Keynesian solution to structural unemployment. For the former University of Minnesota economics professor, the optimum way to mitigate structural unemployment was to enact the proposed Kennedy tax cuts. Heller was certain that when business had money to modernize and expand capacity, and consumers could spend more, economic growth and reduced unemployment would follow. Enactment of the Kennedy tax stimulus, Heller believed, would add two to three million new jobs to an economy that for seventy-two months had experienced unemployment rates consistently above 5 percent.[21]

With Arthur Goldberg having departed to become the newest Supreme Court justice, his replacement, Willard Wirtz, used his appearance at Clark's hearings to convey the same sense of "New Frontier" optimism that the right mix of new policies might alleviate future unemployment. Wirtz was a skilled negotiator and had helped settle major labor-management disputes, such as the longshoremen's strike and the New York newspaper workers' walkout. While in his own career he never wore a blue collar, he was an indefatigable advocate for the unemployed, and he had admitted to an interviewer that he had become emotional about the unemployment problem. As the *New York Times* labor reporter A. H. Raskin observed, Wirtz considered it his number one obligation "to help develop imaginative new solutions for the problems of full employment and manpower retraining in a period when the dizzying pace of technological change often makes him feel that history ought to be arrested for reckless driving."[22]

Before Clark's committee, Wirtz testified that the Department of Labor already had approved training for 38,700 workers in over 1,095 job-training projects nationwide, though of course the department was just getting started. Wirtz proudly reported that the new law was also incentivizing the unemployed to move to job-rich locations, and that Clark's initiative had already jump-started a trend in the employment of those recently retrained and previously unskilled workers. Thus, he noted that it was particularly helpful in moving more African Americans into new jobs. That said, Wirtz saw continuing growth in the skills gap of the US workforce. Exacerbating the problem, retraining no longer meant simply learning new skills for an industrial economy, but also having the flexibility to prepare workers to acquire the clerical and sales skills so much in demand in the emerging service economy.[23]

In the Kennedy administration, the very same differences that defined a now four-decade debate between technology and jobs were often openly, if precariously, balanced. For example, following Wirtz at the witness table, his colleague Commerce Secretary Hodges chose to emphasize how important business growth and economic expansion were to alleviating unemployment. In contrast to Wirtz, Hodges began as a blue-collar worker, quickly moving up the ladder to become the CEO of Carolina Cotton, an example of the booming southern textile firms that Kennedy had described in his maiden floor speeches as a senator. Hodges, like Goldwater, was a politically engaged Sun Belt businessman and had become governor of North Carolina the same year Goldwater became a senator. Though Kennedy valued Hodges as a conduit to conservative segregationist Democrats, much of the governor's legacy was tied to his role in driving the creation of Research Triangle Park. The fast-developing regional technology center that took advantage of three large universities in the Durham-Raleigh corridor served as a national model for the seemingly unbound economic opportunity that resulted from combining government and private research dollars. In this emerging research hub, businesses welcomed government support, creating a business-government partnership far removed from earlier resistance to government assistance. Reflective of an emerging new Sun Belt business ethos, large companies flocked to the region, where they welcomed government dollars to enable their research and development, all the while maintaining their staunch resistance to government interference in other aspects of their business affairs. The business community quickly came to view Hodges as emblematic of the good side of their schizophrenic relationship with the new administration and one of the few friends it had in JFK's cabinet. For his part, Hodges kiddingly described himself to his business allies as representing the administration's "only tie to the nineteenth century."[24]

In his appearance before the Clark subcommittee, the commerce secretary exuded his inherent techno-optimism. He confidently predicted that the gross national product would soon rise from 3.92 percent to 5.2 percent, and then utilized a series of detailed charts and graphs to paint a picture of how technology was accelerating trends toward an urbanized and specialized workforce. Where Wirtz was concerned about the state of vocational education, Hodges worried about the nation's deficiencies in science education. Presciently, he warned that Japan and West Germany, in part due to the United States' own postwar beneficence, were making up ground. Though trade had not yet assumed the status it would in a globalized economy in the not-too-distant future, Hodges was already aggressive in promoting its benefits. Export industries were among those that commanded the highest wages, and thus those with the largest research and development profile. Companies that exported, Hodges asserted, also tended to be the most highly innovative. Like Wirtz, however, he was careful to balance his message. Though technological change was necessary to raise both productivity and incomes, so, too, were policies that minimized unemployment.[25]

Taken together, Wirtz and Hodges, now cochairing the president's Advisory Committee on Labor-Management Policy, provided insight into the administration's expansive view of automation. Though the two reflected a difference in emphasis, they both addressed the president's concern about nagging high rates of unemployment. Wirtz expressed to Congress a general frustration that the relentless pace of technological advancement created a general sense of unease in the public's mind; perhaps, he mused, the steps the department was taking were too little and too late. Automation, after all, seemed to be outpacing policy. Wirtz told the chairman, "Even though we are pressing forward to ever-greater scientific discovery with remarkable success, we do not really comprehend the implications for our society of the current and imminent larger flow of advances in automation and technology— so we have not fully geared ourselves to prepare for the effects of forthcoming changes." Perhaps, then, Wirtz continued, it was time to call forth a commission of national leaders to wrestle with this challenge. Just as the president had suggested several times on the campaign trail, Wirtz mused that perhaps an assemblage of experts might best determine how to allow for technology to continue to accelerate without having to fear its consequences for employment. If forming a commission was simply an excuse for not acting on what was already known, then, the labor secretary said, he would not have endorsed the idea. Rather, he saw the convergence of thought leaders from outside government, just as was the case with the advisory council he cochaired, as the best way to develop innovative solutions and new policies. "We must take advantage both of what they know and of their ability to dramatize this subject in the public mind," Wirtz urged, and Hodges quickly seconded.[26]

The Visionaries

Clark's hearings previewed ideas that such a commission might entertain. For example, one witness, Robert Theobald, a thirty-four-year-old Harvard and Cambridge–educated economist, was alarmed at what he viewed as inadequate job growth and unacceptable levels of unemployment. But he believed it futile to try and stimulate more rapid economic growth through the sort of fiscal measures the administration proposed. It was simply impossible to create enough jobs to outpace what he described as "the juggernaut of scientific and technological change." Instead, Theobald's bold solution was to provide a guaranteed income for all Americans. In an article he wrote for the *Nation*, the young economist distinguished his idea as going well beyond the GAW that Reuther had negotiated with the auto companies. By simply providing Americans with a guaranteed income, Theobald would break the traditional connection between jobs and income. In Theobald's view, the GAW was well intentioned but inadequate given the automated future. Something much bolder was necessary to slow what he described as a "whirling-dervish economy." Theobald was politically astute enough to recognize that his idea extended well beyond the practical horizon of even the most visionary politician. Though already well known for his radical notion, he shrewdly avoided promoting the idea in testimony to Clark's subcommittee. Indeed, it was almost certain that most committee members believed a guaranteed income antithetical to the basic capitalist principle of an incentive-driven market. Theobald instead focused attention on automation's devastating jobs impact. If he could enlist support for the idea that the challenge of unemployment was of greater magnitude than was realized, then perhaps he could lead the skeptical toward his bold solution. If the new automated economy could not provide sufficient and meaningful employment, maybe the only solution was to accept the inevitable and guarantee everyone some basic wage.[27]

Another visionary appearing before the Clark subcommittee provided what at the moment seemed a more politically credible path forward. By now John Diebold's name was synonymous with automation. He had appeared at virtually every congressional hearing addressing the subject over the previous decade. Yet the next development in his thinking, like automation itself, was not always predictable. It was expected that various representatives of the liberal-labor coalition, like Reuther, would continue to call for a federal economic planning capacity that could anticipate automation's next turn. But Diebold's clients read like a who's who of the Fortune 500. It seemed unlikely he would jump onboard. But by 1963, over a decade since the publication of his groundbreaking *Automation* (1952), some of the consequences he had warned about during Patman's 1955 automation hearings were coming home to roost. As he told Clark's subcommittee, "We are not looking far enough ahead." The liberal-labor case for economic planning seemingly had an advocate from an unsuspected quarter. The much-acclaimed business consultant told the committee how the United States' Cold War rival Russia had established a goal of tripling the productive capacity of its computers and then did so in eighteen months. Whether it was accurate or not, Diebold used the story to indicate how the United States, in its research, educational, and employment policies, seemed guided by ad hoc decision-making and chance more than by an astute and carefully coordinated plan. Diebold told Clark's panel that the employment implications of technological change were simply not being grasped and that the consequences would be severe. He agreed that if the government was not willing to establish some better planning mechanism, then the establishment of a presidential commission on automation was perhaps the only way to begin thinking forward.[28]

As the hearings dragged on, it seemed apparent that the liberal-labor coalition, repeatedly frustrated in its efforts to create a stronger central planning and forecasting function, at least was attracting support for forming a presidential commission to devise polices that addressed the automated future. Even before Clark gaveled in his manpower hearings, Oregon senator Wayne Morse wrote a letter to Kennedy urging the president to appoint a commission. Morse grew up as a La Follette progressive in Wisconsin, yet seemed to jettison that upbringing by critiquing New Deal policies long enough to win an Oregon Senate seat in 1944. Once sworn in, however, he gradually gravitated back toward his progressive roots. And though he won again in 1950 as a Republican, he switched parties in 1952. Never doctrinaire in his beliefs or conventional in his methods, Morse perhaps became best known for his later break with President Johnson on the Vietnam War. But in this instance, Morse was convinced from his vantage point chairing a Senate labor subcommittee on education that too many of the nation's youth were not trained for the new automated economy. In his letter to the president, Morse agreed that a commission was now needed to explore how automation would reshape the workforce. "The rapidity with which automated change is affecting adversely both our blue and our white-collar labor forces poses serious problems which must be met," Morse wrote, reenforcing the need for a fresh set of recommendations from a panel of experts.[29]

Eager to get ahead of the curve on automation, Morse was also a cosponsor of legislation introduced in July 1963 by the liberal Republican Jacob Javits that called for the creation of a permanent "National Productivity Council," a national planning body able to forecast and thus mitigate the worst effects of automation. Again revealing the bipartisan nature of the liberal-labor coalition, the Javits-Morse proposal was reminiscent of the very mechanism jettisoned by Congress seventeen years before, during the debate over full employment legislation. In his remarks introducing the bill, Javits critiqued Kennedy's existing Advisory Committee on Labor-Management Policy as lacking the teeth necessary to fight what amounted now to a war with the Soviet Union to increase productivity. The very future of the US economy depended on the United States not only being more productive than any global rival, but also doing so in a way that did not lead to massive worker dislocation. In his remarks, Javits went out of his way to preempt claims from his own more conservative Republican colleagues. He reassured these potential intraparty critics that his proposed National Productivity Council would not have the authority to set wages or prices, nor create production quotas (i.e., it was not a recreated National Recovery Administration). But the council would be empowered to address the shifting employment and production patterns resulting from automation. The Javits-Morse bill called for the council to be made up of labor and management representatives, and those "whose talents and responsibilities would represent all aspects of the U.S. economic complex." Headed by the vice president, the council would also include key cabinet members, and most importantly be a permanent stand-alone office with executive authority.[30]

Kennedy's Call for a Commission

Kennedy had a deep and abiding respect for Javits, and vice versa. The two had arrived in the same class of freshman House members in 1947, and later served together in the Senate. Though they had their policy differences, most publicly when they clashed over a forerunner to Medicare, they worked together on several important pieces of legislation while serving on the Senate Labor Committee. Often socializing together, they became close friends, sharing stories and ideas in the Senate cloakroom. Javits anguished over Kennedy's bad back and delighted in his wit and charm. The first time he saw him as president, Javits greeted him with the formal "Mr. President," and Kennedy responded, "Oh, cut it out." Yet Javits persisted out of respect for their close personal and professional relationship. As Javits would later recall, "He [Kennedy] had a great deal of confidence in my legal judgments as to what would and would not be constitutional, and what particular provisions did and did not mean." Javits possessed a knack for being able to bring conflicting sides together. He later observed that he typically could find the sweet spot where two colleagues—say, the conservative Barry Goldwater and the liberal John Kennedy—could agree. He said that he brought the two together by simply having them acknowledge their differences and then seeking to find common ground. Not surprisingly, then, Kennedy turned to Javits to introduce a bill that would create a Presidential Commission on Automation, even though the New York senator had his own bill to create a permanent commission.[31]

In late July 1963, the opportunity came for Kennedy to formally call for the commission, when he cut short his Cape Cod vacation to return to Washington to deal with the pending nationwide rail strike. On July 22, in a statement to Congress, he plotted a legislative solution to the labor management standoff that threatened to bring the economy to the brink of a shutdown. When Kennedy took office, employment in the rail industry was roughly half of what it was in 1920. The threatened strike turned on several issues, but none more symbolic than the role of the fireman, a position that over time was threatened with elimination. Kennedy's proposal to Congress temporarily saved some ten thousand firemen jobs for a two-year period, or at least until the Interstate Commerce Commission could determine whether to accept, reject, or modify proposed work rules submitted by management.[32]

While Kennedy's message laid out the details of his plan to avoid the rail strike, he also used this opportunity to speak to the broader issues of automation. The threatened strike, with its potentially enormous consequences, seemed a tipping point for Kennedy to formally propose the creation of an automation commission. Thus, at the conclusion of his message, Kennedy reaffirmed his belief in sustaining the United States' impressive technological progress, noting that "this public blessing is not a private curse." Then, echoing those who sought more innovative and longer-term solutions, Kennedy called for the creation of a Presidential Commission on Automation, "composed of the ablest men in public and private life." The panel, Kennedy said, should be empowered by a congressional act and undertake the most comprehensive and innovative review of automation to date. Expressing the sentiments of many liberal-labor reformers dating back to Frances Perkins, he wrote, "Its report must pioneer in the social, political, and economic aspects of automation to the same extent that our science and industry have pioneered its physical aspects." The very language exuded the abiding faith of the "New Frontier" that by assembling the "best and brightest" minds, it was possible to craft policy solutions to the most vexing of social problems. Indeed, the nation's Cold War–inspired technological hubris seemed to spill over into politics. As the contemporary journalist Theodore White put it, evident throughout the Kennedy administration was an amazing sense of confidence "that whatever Americans wished to make happen, would happen."[33]

Javits gladly agreed to formally introduce the president's commission legislation. As proposed, the bill authorized a panel consisting of twenty-five members balanced between industry and labor, science and technology experts, and a contingent of other "public" members. As outlined, the commission's principal task was to do what no analysis to date had been able to do: identify and describe the major technological changes occurring over the next decade and forecast their impacts on employment practices. It would assess the pace of technological change not only in specific industries and occupations, but also in "communities, families, social structure, and human values." The commission was to examine the obstacles to reemployment and address how to ease the transition of those workers displaced by automation. It was also asked to assess the appropriate roles for federal, state, and local governments at all levels. Most importantly, and central to the liberal-labor coalition's long-held goal, the Kennedy bill prescribed that the federal commission forecast technological changes so that policies might be designed to mitigate any near-term negative impacts. As he introduced the bill, Javits noted that the railroad crisis not only resulted from the relentless forces of automation, but also caught everyone flat-footed. "We must not be caught unprepared again on an issue which will clearly be increasingly more critical in every aspect of the American economy," he urged.[34]

The Javits-Morse bill attracted eight cosponsors, including four veterans of prior congressional debates over unemployment: Senators Cooper (R-KY), Clark (D-PA), Douglas (D-IL), and Randolph (D-WV). Rising in support on the floor, Randolph joined Javits in noting that the long-standing problem of technological unemployment had taken on entirely new dimensions. Now in its new automated form, it permeated every aspect of the US economy, including government. The veteran New Dealer was reminded how he had offered legislation in 1939 that would have empowered the secretary of labor to address the impacts of new technology on employment and would have assisted those displaced from work because of scientific advancement. At the time, Randolph reminded his colleagues, those bills did not garner much support. "We have been too long in redirecting our attention to the problems which some of us saw in their embryonic state many years ago," Randolph concluded, as he threw his full support behind the formation of the commission.[35]

In the House, Congressman William Fitts Ryan (D-NY), representing Manhattan's Upper West Side, had earlier offered a bill similar to Javits's original commission bill, proposing the formation of a permanent "Automation Council." In his introductory remarks, Ryan said that whether it was technological displacement or automation, Congress seemed historically incapable of anticipating and preparing for the future. He observed how short-term, stopgap solutions, such as investment in public works, had consistently proved inadequate to address this persistent problem of the industrial era. Yet rather than turn to Ryan to introduce the House version of its bill, the administration tapped James Fulton (R-PA), an early and ardent proponent of both civil rights and women's rights and, like Javits, a well-established, liberal Republican. Fulton had witnessed firsthand technology's impacts on jobs in his Monongahela Valley steel industry district. He agreed that Congress seemed to have exhausted its options, at least those politically realizable. Fulton thus gladly offered the proposal in the House, and the administration was confident that the measure, like most of the presidential commissions formed since President Theodore Roosevelt first made them popular, would pass quickly and without any controversy.[36]

King's Call for Racial Justice

As Kennedy's rail legislation slowly made its way through Congress, the deadline for the rail strike fast approached. On August 28, with rail workers set to go on strike at midnight, Congress moved expeditiously to pass Kennedy's solution. Throughout an afternoon of tense deliberation, members of Congress peered out from the west side of the Capitol at a growing mass of people gathered on the National Mall as part of a march that had been advertised for weeks as "The March for Jobs and Freedom." By six o'clock, just six hours away from the rail union's strike deadline, and with the rail legislation having finally passed, congressional leadership departed Capitol Hill for a quickly assembled White House signing ceremony. The entourage, however, soon encountered the immense traffic and police blockades resulting from the unexpected gathering of hundreds of thousands. Since his threatened 1941 Negro March on Washington, A. Philip Randolph, who had organized the Brotherhood of Sleeping Car Porters in the 1920s, had gained a reputation as an effective advocate for civil rights. And he sought every opportunity to tie the growing civil rights movement to the need for jobs. Though the Negro March never materialized, the very real possibility it would caused FDR to sign Executive Order 8802, the Fair Employment Act, aimed at prohibiting discrimination in hiring in the defense industry. In the immediate aftermath, Randolph worked with the likes of Bayard Rustin to mobilize black support for Senator Murray's Full Employment Act. Structural unemployment disproportionately affected African Americans, and throughout the 1950s an emboldened African American leadership continually voiced concerns about how particularly young African Americans were losing jobs to technology. The nexus between automation, discrimination, and jobs fueled Randolph's desire to rekindle the idea of a march on Washington. On the very afternoon that Congress was preoccupied by its attempt to avert a rail strike, the Reverend Martin Luther King Jr., at the other end of the National Mall, would lift the nation's eyes to promises too long deferred. If freedom were to truly ring, then African Americans must not be deprived of the opportunity for jobs and economic security.[37]

Though rightly seen as a demonstration for fundamental racial equality, beneath the soaring rhetoric of the moment the connection between jobs and freedom resonated. Indeed, Walter Reuther had long understood the connection between racial justice, worker justice, and economic justice. He proudly joined his friend Martin Luther King on the dais at the march, and spoke to the gathering crowd to ask them, "If we can have full employment and full production in pursuit of the negative ends of war, why can't we have a job for every American in pursuit of peace?" Reuther was not some late arriver to the cause. In 1945, before he assumed leadership of the United Auto Workers, he flew to Atlanta to demand that a local chapter admit black members. In 1955 he told his colleagues, "The United States cannot lead the world unless we are ready to fight the master race theory in Mississippi as we fought the master race theory in Germany." Fearing the march might trigger a backlash to its proposed civil rights legislation, one that would serve only to embolden southern Democratic opposition, the Kennedy administration would successfully engage Reuther's influence in the movement to its own political advantage. Reuther's commitment to the cause of racial justice was long and deep. Indeed, in King's very rhetoric, one might hear Reuther's repeated entreaties to management regarding unlocking the benefits of improved productivity so that blue-collar union workers might share in them. "We refuse to believe that there are insufficient funds in the great vaults of opportunity of this nation," King declared as he began his speech.[38]

As the march's ceremonies proceeded on national television, Kennedy's signing legislation that would avert a rail strike was not enough to divert the nation's attention from what was happening on the other side of the South Lawn. The immensity of the March on Washington led the *Washington Post* and other newspapers to make the settlement of the rail strike the second-most-important story of the day. And totally lost in the swirl of events was the fact that although Kennedy called for the formation of an automation commission as part of his original rail solution, it was conspicuously absent from the emergency measure he signed. Javits and Fulton's stand-alone administration-sponsored legislation to form the commission would now idle in each chamber. As fall arrived and with the rail strike averted, it seemed that the idea of forming an automation commission had little push from a White House focused on enacting its tax cut legislation and trying to put through a civil rights bill that would make it illegal for any employer to discriminate against anyone based on "race, color, religion, sex, or national origin"—precisely the language that would be enacted a little over a year to the date after Kennedy introduced it.[39]

Though forming a commission now seemed at best a secondary legislative priority, the concern over automation continued. On November 9, two weeks before the president was scheduled to go to Dallas, Wright Patman, whose hearings in 1955 had formed an important record on the then-little-understood phenomenon of automation, dutifully addressed a 4-H Club in Mount Pleasant, Texas. Patman described how, even in rural communities, and particularly on the farm, automation was reshaping the future workforce, and he now demanded that as a nation, the United States respond with forward-thinking solutions. More concerned than he was at the time of his hearings, Patman warned that "automation has no time, no place, no patience with the 'standpatters'—the people who want to go along with things as they are, or as they used to be. In the real world of today, there are no wagon trains to the past." It seemed clear that seventeen years after such a planning mechanism was stripped from the Full Employment Act, many in Congress were reassessing the liberal-labor desire that government become better prepared to determine technology's next turn.[40]

Similarly, other members were not waiting for a commission to be formed, as they offered new approaches that they hoped would recast technology so it might better serve societal needs. Among the new liberal-labor reformers was the first-term senator Philip Hart (D-MI), who was already gaining a reputation for being an effective, though low-key, sponsor for major civil rights, consumer, and antitrust bills that would be among the decade's many significant legislative achievements. Hart would eventually become known as "the Conscience of the Senate." On November 8, 1963, he offered his somewhat different approach to a Presidential Commission on Automation. He suggested that the federal government itself was enabling the automation problem. The government's now-prodigious research and development budget had grown some 600 percent since 1953, with the vast majority going to sustain the United States' competitiveness against Russia in the Cold War. Hart proposed an automation commission that would explore whether the return might be greater if government directed the focus of technology on unmet community needs in health care, transportation, and poverty. Using the pending debate over government funding of a supersonic bomber as an example of the misappropriation of funds, Hart asked, "What could be accomplished if an equal amount of public research and development money was committed to community needs?" His proposed commission would identify specific areas where government-funded technological expertise might be applied to community problems. The proposed "Commission on the Application of Technology to Community and Manpower Needs" quickly won the cosponsorship of Hart's fellow liberal senators Humphrey and Clark. While it added momentum to the larger need for an automation commission, it also reflected a new attitude within the liberal-labor coalition. While Kennedy's expansive view of technology seemed still the predominant Cold War perspective, even as Kennedy sent the first "advisers" into Vietnam, new voices like Hart's were daring to question the Pentagon's disproportionate influence on the course of technological development.[41]

Johnson Takes the Reins

While Senator Hart foreshadowed a coming backlash against the onslaught of technological progress, union leadership was renewing its own warnings. The week after Hart introduced his own commission bill, and a week before President Kennedy left for Dallas, George Meany stopped chewing on his trademark cigar long enough to give the presidential address at the AFL-CIO's annual meeting. From the lectern at the Americana Hotel in New York City, Meany bolstered his own patriotic credentials by praising Kennedy's realistic and strong defense policies. When he turned to automation, the labor boss acknowledged that some of the experts reassured the working man that new technology would mean new jobs. Even if that were true, Meany suggested, what should laborers expect in the near term? "Do we get our purchasing power from the push-buttons?" he said. Meany thought the record on automation clear. He claimed that an estimated four million jobs would be lost to technology and automation over the next year, and he angrily challenged an earlier speaker, New York governor Nelson Rockefeller, who stated that automation was a mixed blessing. "There is no element of blessing," Meany retorted. "It is rapidly becoming a real curse to this society. When you study what is happening, you realize this is a real threat; this could bring us to a national catastrophe. Every big corporation in America is in a mad race to produce more and more with less and less labor without any feeling as to what it may mean to the whole national economy." Warning that corporate America was not about to back away from producing more products with fewer workers, Meany predicted that companies would soon be building more "push-button operations with machines to push the buttons." The system was, as Meany bluntly put it, "about to go down the drain." Something needed to be done about automation, even if it was simply authorizing a presidential commission.[42]

The week after Meany's rambunctious speech, one of the first phone calls that Johnson made as the nation's thirty-sixth president was not to Meany, but rather to his friend Walter Reuther. The newly sworn-in president, his tone of voice reflecting the pall of a grieving nation, said plainly, "Walter, I need your help." Though bruised by the investigative hearings that Senator McClellan held in 1960 that spotlighted labor's ties to organized crime, the unions remained essential to the liberal-labor coalition and thus to achieving the fallen President Kennedy's stalled legislative agenda. Johnson knew how many votes labor meant, not only at the ballot box, but in the halls of Congress. When the new president called to summon his help, Reuther promised Johnson he would give it. As the new administration took center stage, it was not uncommon for Reuther to quietly find his way to the Oval Office and spend a couple of hours strategizing and brainstorming with the president. As Humphrey would later say, "He [Reuther] was a practical man, and there are very few liberals who knew how far to go." No wonder, then, that he and the pragmatic new president got along famously. On the subject of automation, one did not have to strain to see Reuther's influence on Johnson. Not only would Johnson agree that an automation commission was necessary, but he would cast his request using Reuther's familiar refrain, by urging Congress to be as innovative with policy as business was with technology.[43]

Throughout his political career, from his first days as the secretary to Congressman Richard Kleberg (D-TX), Johnson seemed always in a hurry to do big things. On January 8, 1964, as he came before Congress to deliver his first State of the Union address, a mere forty-seven days after Kennedy's assassination, he was dutifully aware of the political necessity of paying respect to the slain president's memory. Still surrounded by most of the former president's appointees, Johnson knew that before he could turn the ship of state in his own direction, he first needed to honor the goals that Kennedy had set—even if they were objectives that were established by Kennedy people who seldom sought, and just as often ignored, the vice president's input. Johnson would use the speech to invigorate a moribund yet ambitious congressional agenda, calling for Congress to act on the unfinished business before it, from tax cuts to civil rights and, yes, even a Commission on Automation.

Members on both sides of the aisle welcomed their old friend with a thunderous ovation as he slowly loped toward the dais, accompanied by House and Senate leadership. Having addressed a joint session of Congress the week after the assassination, the president wasted no time getting to the point in his January address. "I will be brief, for our time is necessarily short and our agenda is already long," he said. Then, in his soothing and sonorous Texas drawl, he began to reel off the contours of his ambitious agenda, including a War on Poverty, the expansion of health care to the elderly, and finally making good on the century-old promises of civil rights. Each initiative received hearty applause from both parties. Virtually no constituency in the United States was left off the president's to-do list. The sense of energy Kennedy had exuded around his lofty goals now had the full force of Johnson's immense political skills. As he proceeded, Johnson reassured his critics that these goals could all be accomplished while reducing the federal budget and the number of federal employees. Then, among a litany of "we musts," the president said, "We must modernize our unemployment insurance and establish a high-level commission on automation. If we have the brain power to invent these machines, we have the brain power to make certain that they are a boon and not a bane to humanity." Congressional action to create a blue-ribbon panel to study the vexing phenomenon of structural unemployment may not have been the new president's first priority, but by adding it to his long list of programs and proposals, Johnson signaled that the formation of a commission was not a matter of whether it would be authorized by Congress, but of how quickly.[44]

The 306 days remaining before the 1964 election would be marked by a flurry of bills enacted and signed into law by an energetic new president, including one to create the National Commission on Technology, Automation, and Economic Progress. After almost four decades of chasing after the effects of new workplace technologies on unsuspecting employees, Congress and the administration seemed to throw up their hands. Perhaps a panel free from the immediate pressures of everyday politics could devise the sort of innovative policies that would ameliorate the short-term impacts of technological development. At the same time, the creation of the commission evinced the hubristic attitude that in the United States, no problem was too big to solve. If nothing more, such a collection of experts might at last reinvigorate the liberal-labor vision of a federal planning mechanism that could blunt the worst effects of an increasingly automated workplace and anticipate technology's impacts on innocent workers.

CHAPTER 7

The Automation Commission

Automation is not our enemy. Our enemies are ignorance, indifference, and inertia.

—President Lyndon B. Johnson, remarks at bill signing for the National Commission on Technology, Automation, and Economic Progress, August 19, 1964

Five days after President Johnson's State of the Union address, Norbert Wiener and Vannevar Bush gathered in the Cabinet Room of the White House to receive the National Medal of Science. Congratulating all five of the recipients on their outstanding contributions, LBJ seemed to have a particularly warm regard for Bush. Indeed, in his role as Senate majority leader, just after the launch of Sputnik in 1957, Johnson welcomed Bush's forceful call for a national space program. Bush's advocacy lent credibility to the legislative response to Sputnik that Johnson skillfully maneuvered through Congress: the creation of the National Aeronautics and Space Administration (NASA). Johnson would remain tethered to the space program throughout the rest of his political career. For the master legislator who knew firsthand the ills of poverty, there was always a corresponding confidence in the potential of technology. Johnson believed that technological advancement was critical to winning the Cold War and boosting standards of living for all Americans. The nation's "complex fabric," Johnson noted in his remarks at the ceremony, demanded the nation's continuing confidence in the promise of science and technology.[1]

During the White House ceremony and in conferring the award on Wiener, Johnson praised Wiener's discoveries as both "marvelously versatile" and "profoundly original." Wiener, however, continued to harbor reservations about the consequences of automation. Recently he had become more determined than ever to turn cybernetics into a force for good. From his own experience as an enlistee in World War I, he recalled seeing the horrific sight of dismembered veterans returning home. Inspired by a trip to Russia, where he witnessed firsthand how his language of cybernetics was being used to create new prosthetic devices, he decided to advance bionic science at home. With a team of Harvard doctors, Wiener helped develop a prosthetic hand called the Boston arm. Though Harvard initially received more of the credit for this technological breakthrough, Johnson acknowledged that Wiener had "penetrated boldly into the engineering and biological sciences." As honored as Wiener was to accept the medal from the president, those of his friends attending the ceremony noted how tired and visibly overweight he was. A few weeks later, an announcement came from Cambridge, Massachusetts, announcing that Wiener had died of a heart attack in Stockholm, where he was scheduled to give a lecture. As much as anyone's, Wiener's insights from three decades before were now fueling an automation revolution that few had seen coming and that challenged politicians and policy experts to be just as creative in their response. Wiener and his longtime MIT colleague Vannevar Bush, standing together in the Cabinet Room, embodied the larger dimensions of technology's oversized imprint on the US economy: Wiener the sheer force of unpredictable scientific discovery, and Bush the opportunity to manage science in pursuit of larger societal goals. Johnson, still preoccupied with the challenges of his sudden transition, now hoped that his predecessor's idea of forming an automation commission would create a set of policy recommendations every bit as imaginative as Wiener's discoveries, and harness technology's impacts on employment in a manner befitting of Bush's managerial brilliance.[2]

Legislating a Commission

Two days after the White House ceremony honoring Bush and Wiener, Hubert Humphrey made his way to the Senate chamber to offer his own version of an automation commission. His loss to Kennedy in the hard-fought 1960 Democratic presidential primaries had done nothing to dim the Minnesota senator's characteristic enthusiasm. Humphrey's ebullience and ingrained optimism belied the fact that he came of age during the Depression. Indeed, he seemed most passionate when addressing issues of discrimination, poverty, and unemployment. One of his aides recalled how Humphrey's office often served as a surrogate employment agency. For example, at the request of an imprisoned man's wife, he used his senatorial privilege to move her convicted husband from a jail in Kansas to one nearer the couple's Minnesota home. When the felon was released, Humphrey helped find him another job at a slaughterhouse. And when the slaughterhouse closed, Humphrey secured him yet another job as a truck driver and followed up by writing him a congratulatory letter. Not only was Humphrey committed to policies that helped the unemployed, like the new president, he personally connected with those the economy left behind.[3]

Asking for recognition from the presiding officer in the Senate on the afternoon of January 15, 1964, the cherubic-faced senator proceeded to explain why he was offering legislation to form a Presidential Commission on Automation. "It is time to stop wringing our hands about present problems associated with automation and technology, and the large ones looming. It is time to start attacking them," Humphrey urged. Though his commission bill differed from what the president would send to Congress two months later, he borrowed from his mentor's inclusive approach to politics by praising Senators Javits and Hart for their own versions of similar legislation. Soon to be at the forefront of guiding the administration's landmark civil rights legislation, Humphrey used this occasion to reinforce how technological development might serve to improve depressed neighborhood conditions, and he stated his desire that the commission he was proposing might tackle this challenge. To be sure, Humphrey's commission bill was expansive in both in its charter and its structure. The proposed thirty-two-member commission was modeled after the bipartisan and consensual postwar Hoover Commission that sought to remodel the federal government. Regardless of the political stripe of its namesake, Humphrey was enamored with the Hoover Commission's balanced political structure and extensive engagement on a range of topics.[4]

With his own far-ranging version of an automation commission, Humphrey had moved ahead of the new Johnson administration, which was still undergoing a dizzying transition. The new president was attempting to keep on his own team the requisite number of Kennedy men and stay loyal to the extensive New Frontier agenda. Nonetheless, fault lines emerged within the transitioning White House, and even the straightforward task of drafting language for something as politically uncontroversial as creating a presidential commission could become the source of internal dispute. The Kennedy holdover Larry O'Brien, whom Johnson pleaded with to stay on to shepherd the tax and civil rights legislation, placed the staff member Sam Hughes in charge of the commission bill. In later promoting Hughes, Johnson would refer to him as "one of those quiet but highly effective civil servants." Vetting the commission bill through relevant departments, Hughes told the president's counsel, Myer Feldman, that some minor modifications were made concerning the commission's structure. As it turned out, the consultative process had caused a much greater stir. Secretary of Commerce Hodges felt that his department had been completely left out of the process dominated by Secretary Wirtz. It would take a call from LBJ himself to settle Hodges down. Despite the internal snafu, the bill emerged from the White House process expeditiously, and by March 9 a revised draft bill was ready for O'Brien and his team to navigate through Congress.[5]

In his accompanying transmission message, Johnson underscored his abiding faith in technology while also citing his continued concerns over unemployment. "We must make sure that as technological progress creates new industries and job opportunities it does not impose too great a hardship on individual workers," he wrote. This language echoed that used earlier by labor leaders such as Green and then Reuther, and embodied a theme that business spokesmen such as Watson and Diebold embraced. For both labor and business, over the span of three decades, the rhetorical challenge was to strike a balance between the nation's belief in technological progress and its concern for the collateral human toll. The political challenge remained how to achieve such a balance when technology continued to race ahead of politics. Just like his predecessor, Johnson hoped a blue-ribbon commission might finally recommend methods and policies that could lead to reconciliation between these two important principles. And a commission could perhaps, at long last, give credence to the liberal-labor coalition's desire for an economic planning mechanism.[6]

The Johnson Administration's commission legislation was introduced in the House by Representative Elmer Holland (D-PA) and in the Senate by Philip Hart. Like the identical bills offered by Fulton and Javits the year before, the administration's bill prioritized the need for more accurate forecasting of those industries, occupations, and regions where technological advances were having the most impact. The bill thus required the commission to measure the pace of technological change. It also borrowed from the approach taken in both Humphrey's and Hart's earlier commission proposals by requiring recommendations on how technological advances could be applied to large-scale human and community needs. And the bill further tasked the commission with recommending specific initiatives that management and labor should jointly undertake to spur economic growth while helping to prevent and alleviate the adverse impacts of technological change, in particular the displacement of workers. Not only did the bill map a bold legislative charter, but it also required a final commission report within a year of enactment.[7]

Sounding the Alarm

The commission bill was put forward by the administration as the always-shifting tectonic plates of US politics were about to undergo their own seismic realignment. The seeming post–World War II political stasis was not quite as buttoned-down as the nation's lily-white suburbs or its ivy-adorned campuses conveyed. The Freedom Riders (1961), the Port Huron Statement (1962), Martin Luther King's letter from the Birmingham jail (1963), and the Berkeley Free Speech movement (1964) had already been early indications of the coming social turmoil about to be exacerbated by LBJ's escalation of the Vietnam War. Indicative of a growing new reformist impulse, two weeks after the introduction of the administration's commission bill, the White House received a memorandum with an accompanying cover letter signed by twenty-six members from the Ad Hoc Committee on the Triple Revolution (AHC). Addressed to the president, the letter began by dramatically warning that if the policies it outlined in an attached fifteen-page memorandum were not soon adopted, "the nation will be thrown into unprecedented economic and social disorder." As the political commentator Hallock Hoffman later noted in an article in the *North American Review*, the memorandum "set voices and typewriters and editorial pages ringing all over the country." The AHC included Robert Theobald, the futurist and strong advocate of a guaranteed income; Todd Gitlin, the antinuclear activist; and Tom Hayden, head of the Students for a Democratic Society, who had earlier authored the Port Huron Statement for social and economic justice. The AHC also included one of Kennedy's and Johnson's favorite authors, Michael Harrington, whose book *The Other America* gripped the nation following its publication in 1962. Freely exercising his politics of inclusion, Johnson invited Harrington to take a post in the Department of Labor, where the political science professor could help craft the administration's War on Poverty. "If automation continues to inflict more and more penalties on the unskilled and the semiskilled, it could have the impact of permanently increasing the population of the other America," Harrington concluded. In the midst of mounting signs of social unrest, Johnson began to join the concepts of poverty and unemployment, and the AHC's memo reinforced why the two needed joining.[8]

The AHC's eclectic and distinguished membership focused on three major revolutions that it claimed were occurring simultaneously: the threat of nuclear holocaust, discrimination against African Americans, and what it termed the "cybernation revolution," a tribute to Wiener's impact. The AHC claimed these convulsive developments were collectively reordering existing values and institutions. In its focus on automation, it noted that the use of computers to run large factories was posing more than just an existential threat to full employment; it was causing a complete realignment of the economic and social system. The AHC acknowledged that although cybernation possessed the potential to create an abundance of goods, it also produced growing numbers dependent on government programs, cut off from the dignity of meaningful jobs and their share of a fairer distribution of wealth. A new economic paradigm was required, tied not to jobs but to the quality of human life. Drawing directly from Theobald's latest book, *Free Men and Free Markets*, the AHC argued that rather than struggle against the inevitability of an automated workplace, the federal government should accept its inevitability and make "an unqualified commitment to provide every individual and every family with an adequate income as a matter of right."[9]

The memorandum garnered such significant public attention that the White House felt compelled to issue an immediate response. Though not dismissive, its three-paragraph letter attempted to reassure the AHC that the president was addressing each of the three revolutions cited in the memorandum. Despite Johnson's tactful reply, the idea of a basic income drew sharp critiques from a number of establishment quarters. The *Baltimore Sun* parodied the report as "Ad Hokum." The immediate need, it editorialized, was to find new jobs for the unemployed, not give everyone a handout. Redesigning the economic system needed to wait for another day. While the *Washington Post* agreed with the committee's concerns regarding the arms buildup and the civil rights struggle, it found the obsession with automation overdone. The *Post* suggested the committee "should have explored the psychological and sociological consequences of divorcing income from effort." As the criticisms mounted, Wirtz chimed in, noting that a guaranteed income was the wrong answer. He advocated revitalizing skills-based education for American youth as the single most important step that could be taken to confront the changing complexion of jobs in an automated society. But he even went so far as to propose that the first two years of college be made available free to all eligible students.[10]

Despite the establishment backlash, the debate over the ideas raised in the AHC report continued to simmer. The AHC member and Syracuse University professor Michael D. Reagan pushed back against the critics in a highly readable summary of the controversial memorandum for the *New York Times Magazine*. Reagan's article made the case for a guaranteed income by effectively rebutting all the arguments being made against it. For example, Reagan freely acknowledged that the idea stood in sharp defiance of the Protestant ethic and its emphasis on return for hard work. Take away the incentive for work and who would do the most menial of jobs? Reagan answered his own question by suggesting that if these jobs would pay more in the first place, perhaps more workers would be incentivized to do them. At any rate, he expected that most laborers worked at jobs that in the long run would be eliminated by machines, eventually freeing all humans from such demeaning chores. As for the related argument that guaranteeing income caused more indolence and selfishness, Reagan suggested that society was full of creative young people who might make a choice to use their guaranteed income to give back to society, through endeavors ranging from the artistic to the scientific. If cybernation was eliminating jobs faster than new ones could be created, why not dignify each individual's worth by simply moving ahead and granting a basic income? If automation kept driving more individuals to the unemployment rolls, Reagan argued, inevitably we would have no other choice.[11]

The Congressional Hearings

The Triple Revolution memorandum would be made part of the hearing record when Holland's Select Subcommittee on Labor met to review the administration's bill to create an automation commission. But Chairman Holland had no interest in a prolonged hearing or in giving the AHC a platform. Serving his fifth term, he was a veteran of both world wars and had a reputation for combativeness. During his earlier time as a state legislator, he had physically confronted a reporter in the press gallery. A staunch union supporter, he was born, raised, and educated in the Irish district of Pittsburgh, and his legislative career was marked by his loyalty to his blue-collar constituents. Kennedy had leaned heavily on the feisty congressman to steer the Manpower Development and Training Act through the House as the companion to Douglas's Senate bill, and now Johnson turned to Holland to shepherd the commission bill. The chairman saw no reason to invite too much scrutiny over this noncontroversial measure, and he limited his brief April 14 hearing to testimony from Secretaries Wirtz and Hodges.[12]

While limiting witnesses, Holland did invite a range of interested parties to submit statements. For its part, the AHC restated its dire predictions about an automated future, while agreeing that a commission would be worthwhile. Reuther used the opportunity to advance a new UAW position, calling for the establishment of a "Technological Clearinghouse" that would examine all aspects of how new technologies were impacting the economy. "If we are to gain the fullest advantage for advancing technology in this age of automation," he wrote, "we need far more information about what is taking place in the field of technology, what it is doing to our economy, and what it is likely to do in the near future." Speaking for organized labor, Reuther, though endorsing the administration bill, similarly urged the formation of a permanent commission, like Javits's Productivity Council, that could provide continuity in managing automation. While he welcomed and would serve on the president's commission, Reuther would continue to argue for some, indeed any, form of government planning mechanism with enforcement power. In his view, the commission was the best hope for reigniting this long-overdue liberal-labor objective. Two other future appointees to the commission agreed that the commission's time had come. The new CEO of IBM, Thomas Watson Jr., reiterated his father's earlier-expressed sentiment that better policies needed to be developed to address the hardships automation imposed. "We have had a chronic unemployment rate of between 5 and 6 percent since 1957," he said, and it was time that policies adjusted to the new reality. Communications Workers of America president Joseph A. Beirne, who would soon be named to the commission, agreed that the consistently high levels of unemployment required a fresh look by a body free of political pressure.[13]

In the brief hearing Holland conducted, the most forceful questions for Wirtz came from Congressman Clarence Brown, who would leave his hospital room in Bethesda the following summer to help shepherd the Voting Rights Act of 1965 through the Rules Committee. A successful publisher, the youngest lieutenant governor in Ohio history, and a staunch conservative supporter of his friend Senator Robert Taft, Brown was recognized for his courageous stands against lynching and in favor of civil rights legislation. At the same time, he was committed to leaner, more efficient government, and he thought the proposed commission was redundant and wasteful. Reflecting some sympathy with liberal-labor goals, he remarked to Wirtz, "We may be studying this situation to death and not moving toward action fast enough." But Wirtz did not help his cause when he could not answer Brown or other Republicans regarding how much should be budgeted for the commission. In one dissembling response, the Secretary said that "the range of financial possibilities is almost infinite." It was not a reassuring answer to Republican critics.[14]

Yet Brown and Wirtz were able to find some common ground. Brown was interested in how the new generation of IBM computers might make it possible to apply the new science of "systems analysis" to render a more complete understanding of the impact of technology on job growth. Importantly, Brown was interested in the rate of "technological adoption," industry by industry. It was the same question that had prompted the analysis conducted by Ewan Clague at the outset of the Depression, and near its end by Isador Lubin for the TNEC. It was the conundrum that Kennedy's Advisory Committee on Labor-Management Policy had just wrestled with, and now, yet again, it was to become a central feature of a blue-ribbon commission. Indeed, the legislation required the commission to engage in what was hoped would be an even more definitive analysis that might provide insight into the pace of technological change and its impacts on future employment. In response to Brown, however, Wirtz noted that in the department's recent analysis for the Labor-Management panel, in which it studied thirty-six separate industry categories and measured technology's impacts on employment, fourteen had enhanced their computerization capacity. The result, Wirtz pointed out, was an expansion of jobs rather than a contraction. Unfortunately, there were sixteen other industries where the employment picture seemed much bleaker. The report's projections were, Wirtz acknowledged, preliminary and did not warrant any generalization as to whether employment would be going up or down. Yet there were some industries where Wirtz felt somewhat confident in predicting a jobs increase. What troubled Brown about the Labor Department's analysis was that it was unclear as to whether there could be any reasonable projection of a net job gain or loss in the steel and auto industries so vital to the economy of his North Central Ohio district. "I am a little surprised," Brown told Wirtz, "that you cannot make projections in iron and steel and motor vehicles." Testifying the following day, and alert to Brown's concerns, Commerce Secretary Hodges asked rhetorically, "Would it not be in the national interest to develop a more extensive early warning system which would signal the introduction of new technologies and of whole new industries, especially if they replace established ones?" In assessing "the prospective role and pace of technological change," the Automation Commission would eventually devote much of its analysis to deciphering the so-called diffusion rate of technology into the economy, devoting an entire appendix to it. For the liberal-labor coalition, conducting such an analysis was at the foundation of planning intelligently for the future, though previous analyses to date had seemed quite adequate in the face of innovation and invention.[15]

Meanwhile, on the Senate side, Senator Joseph Clark's Subcommittee on Employment and Manpower used part of its July 6 hearings on his favorite subject of manpower training to address the commission legislation. As in the House, Wirtz and Hodges were again the only witnesses, and Clark himself seemed less than enthusiastic about the need for a commission. He had just spent months exhaustively exploring all aspects of the unemployment question and the issues surrounding automation, and he felt that all the policy options had already been fully examined. Wirtz tried to reassure Clark that a "high-level panel" would help focus the nation's attention as no committee or department could. Nonetheless, Clark remained skeptical that any commission, no matter how high powered, could reveal any new policy solutions that might allow policy to get ahead of the arc of accelerating technology. Frustrated, Clark thought it naive to think that after all the consideration already given to technology in the workplace, a presidential commission could develop any new solutions. He asked Wirtz, "Don't you feel in your own experience an accelerating rate of new developments in this whole technological field, and don't you share my fear that the social sciences will not be able to keep up with the enormously accelerated progress which appears to be being made in [the] national physical sciences every day?" It was of course a rhetorical question, to which Wirtz obligingly agreed. Perhaps at this juncture in the now four-decade effort to address the impacts of technology on employment, the call for a commission was itself a tacit admission that the political system was neither fast enough nor creative enough.[16]

The Commission Takes Shape

As the noncontroversial commission legislation emerged from both the House and Senate committees, despite efforts by the Chamber of Commerce to prescribe an even balance between labor and management appointees, the bills as reported left it entirely to the president to appoint the commission's fourteen members, with no specification as to affiliation. The commission was to operate on a $2 million budget, with results to be delivered by January 1, 1966. The July 21 floor proceedings on Congressman Holland's H.R. 10310 were largely a formality. Member after member followed Holland to the well in support of the bill, although seventy-five members, including thirty-two southern Democrats, would eventually vote against it as unnecessary and wasteful. By contrast, the Senate bill passed by unanimous consent after slight modifications, including reducing the operating budget to $1 million, requiring the president (rather than the commissioners) to name the executive secretary, and adding a representative of the Arms Control and Disarmament Committee to the administration's advisory committee. The week after the Senate vote, the House agreed to these minor Senate amendments on a voice vote, and the bill moved to the president for signature.[17]

By the time the enrolled bill reached the White House, staff found themselves scurrying to invite some fifty prominent business, labor, and aca-
demic leaders to the White House for the signing ceremony. On August 19,
1964, President Johnson ambled into the stately Cabinet Room down the
hall from the Oval Office. There, flanked by a hastily assembled gathering
of about a dozen or so union and business leaders, he signed the bill formally establishing the National Commission on Technology, Automation,
and Economic Progress. When Johnson first arrived in Washington in 1931
as an aide to Congressman Kleberg, and with the country in the midst of
the Great Depression, much of the nation's unemployment was thought to
be attributable to the accelerating advance of new technology. Now, thirty-
three years later, and with Johnson only a week away from being nominated
in his own right as his party's choice for president of the United States, technology seemed the cause of persistent structural unemployment. Amid the
many challenges he faced, Johnson seemed content to delegate this nagging
problem to a commission he appointed. His friend and informal adviser, the
columnist Eliot Janeway, wrote that seething beneath the surface of racial
tension was a bleak reality: though the president had earlier signed the landmark Civil Rights Act of 1964, Janeway predicted that automation, coupled
with an economic downturn, might soon result in northern white working
men "lashing back" against their black counterparts as plants shuttered and
jobs were lost. Though the unemployment rate was going down, Johnson
now turned to the appointment of fourteen individuals who might offer new
solutions that could perhaps head off any such crisis.[18]

At the signing ceremony, looking out over his reading glasses, the president told those gathered that the nation needed to attend to the casualties of
rapidly advancing technology. Johnson emphasized that technological progress need not come at the expense of the working man, and he reiterated his
belief that technological progress and the desire to see full employment were
both attainable goals. He confidently stated, "Technology is creating both
new opportunities and new obligations for us—opportunity for greater productivity and progress—obligation to be sure that no workingman, no family must pay an unjust price for progress." Trying to strike the appropriate
balance between technology and jobs, and with his calming and measured
Texas drawl, the president reassured those gathered, "If we understand it,
if we plan for it, if we apply it well, automation will not be a job destroyer
or a family displaced. Instead, it can remove dullness from the work of man
and provide him with more than man has ever had before." Johnson firmly
believed that a Great Society could straddle these two abiding American
beliefs: one based on technology's promise to spur economic growth and

greater wealth, and the other based on the long-held conviction that any individual who wanted to work ought to be able to find a job. Perhaps a commission could finally develop policies that would reduce the recurring conflict between those extolling the job-creating virtues of automation and those who warned that the automated economy would simply continue to grow the unemployment rolls.[19]

For the hastily recruited businessmen present for the signing, the very wording used to describe the commission reinforced their growing comfort level with the new president. While Johnson was still viewed as more pro-labor than pro-business, the leading figures in the business community were by and large supporting him against Goldwater in an election now just three months away. Earlier in the year, Johnson had surprised members of the prestigious Business Council by inviting these influential CEOs to the White House for dinner. One of them noted that not even Eisenhower had done that. At this early juncture in the new Johnson administration, business leaders saw in Johnson a pragmatic politician running on a growing list of legislative accomplishments. While they readily conceded that Reuther and other labor leaders might have special access to the president, they nonetheless welcomed the formation of the new Automation Commission and Johnson's recognition that continued technological advancement was critical to continued economic growth.[20]

For the labor representatives witnessing the signing, the broad charter of the new commission would allow ample opportunity for them to argue their case as well. They saw in Johnson a sympathetic ally who might help them share in the rewards of increasing corporate productivity and profitability. As Reuther himself had urged a decade earlier, policy advances needed to match the pace of rapid technological progress. If automation was creating more productive enterprises and more jobs, as its industry proponents claimed, then it needed to be accompanied by policies that provided a fair share of the returns to the working man. "The buying power of wages and salaries in this period must increase faster than the national economy's rising productivity," Reuther concluded.[21]

Johnson's brief statement at the signing ceremony was indicative of just how focused he was on the upcoming election. His political instincts, always on high alert, were in clear evidence even in these ceremonial remarks. Not only did he note that the unemployment rate was down, but he identified the stimulus of the new tax bill (i.e., the Kennedy tax cuts) that he had signed in February as the reason that GDP was growing at well over 5 percent. Johnson thus took the opportunity to praise business for doing its part to create new jobs as it installed modernized equipment made possible by the cut in

the capital gains tax. He confided how one executive told him that the enactment of the recently passed Kennedy tax legislation would mean twenty
thousand more jobs. Yet as evidence of his big-tent political philosophy, the
very next day, in a far more elaborate signing ceremony, Johnson signed the
Economic Opportunity Act of 1964, effectively launching his War on Poverty that would begin to broaden the unemployment debate. He had already
begun to draw the connection between poverty and joblessness, stating that
"very often the lack of money is not the cause of poverty, but the symptom."
For Johnson, the business of building his inclusive Great Society was taking
on new dimensions as the November election approached. With the White
House already receiving suggestions on whom to nominate to be on the
commission, Johnson would find it politically advantageous to disappoint no
one. He would wait until after the election to name its members.[22]

Appointing the Commission

Throughout the fall of 1964, the White House received dozens of letters
suggesting candidates for the new commission. Yet even those in a seemingly powerful position to influence Johnson's final selection fared poorly.
The chief lobbyist for the American Bankers Association, Charles E. Walker,
was already cultivating a reputation in Washington as one of its most powerful influence peddlers. Enjoying his lunches at the fashionable Sans Souci
restaurant, the tall, cigar-smoking, boisterous, and affable Texan was at the
outset of a career that would eventually make him one of Washington's
most influential corporate tax lobbyists. In 1965, already benefiting from his
extensive Texas connections, Walker conveyed to Johnson special assistant
Walter Jenkins the names of five prominent bankers as potential candidates
to serve on the commission. Walker did not know that Jenkins was already
under investigation for a sexual encounter that, in the context of a homopho-
bic era, would quickly lead to his dismissal from the White House prior to
the November elections. None of Walker's recommendations would make
it to the final list of the president's nominees. Moreover, Johnson would not
appoint a single banker to the commission, even though banking was a business sector whose clerical functions were clearly impacted by automation.
But Walker turned out to be hardly alone among those influential voices that
struck out with their recommendations.[23]

    The White House's vetting of candidates was coordinated by Walter
Heller, the chairman of the Council of Economic Advisers. Filtering the
nominations was Ralph Dungan, a Kennedy holdover whom LBJ would later
appoint to be ambassador to Chile. Secretary Hodges bristled at an internal

memo that characterized a Commerce draft of recommended nominees as "final." In the end, however, five of Commerce's recommended nominees were selected, and Secretary Wirtz was similarly successful in gaining the requisite balance of union representatives. Less successful were members of Congress. John Macy, in the legislative affairs shop, was responsible for shepherding the congressional suggestions, making sure to inform Heller of those individuals recommended by no fewer than six senators and four congressmen, including Speaker John McCormack. Senator Hart himself would recommend thirteen candidates, none of whom were selected. Meanwhile, Elmer Holland, who had steered the authorizing legislation through the House, was successful only in his recommendation of Robert Ryan, who was then serving as president of the Regional Industrial Development Corporation of Southwestern Pennsylvania. Ryan was by far the least known of those eventually chosen, and the only one without a national reputation. Though outspoken on the effects of automation in the Pittsburgh region, Ryan was perhaps just as surprised at the appointment as everyone else was, and graciously acknowledged that he would provide a regional perspective not possessed by the other appointees with broader national cachet. At the same time, John Diebold, arguably the most identifiable national expert on automation, though endorsed by Senators Humphrey, Clark, and Javits, failed to make the final list.[24]

While the administration decided to delay the formal announcement of commission members until after the election, Johnson seemed to have already made up his mind about who should serve as chair. On October 30, Stan Ruttenberg, a former AFL-CIO research director then serving as an assistant secretary of labor, called John Clinton to arrange a call with Wirtz. Ruttenberg suggested that Wirtz would want to discuss with LBJ the selection of the chairman's executive secretary. Apparently, Johnson had quietly tapped Howard Bowen, president of the University of Iowa, to serve as the commission chairman, and Bowen was already engaged in a search for the right executive secretary who would staff the commission. Wirtz had an internal candidate to recommend, and he would reach out to the president to make his case.[25]

Bowen became president of what was known as the State University of Iowa in 1964, after a successful tenure at nearby Grinnell College, where he landed after being forced out of a deanship of the business school at the University of Illinois in 1950. In a much-publicized controversy, Bowen was accused by fellow faculty of being too "anti-business." Helping to fuel the critique, the conservative *Chicago Tribune* offered that it hoped that no more "disciples of the Keynes doctrine will be added to the Illinois faculty." The

Bowen Affair, as the *Tribune* disparagingly referred to it, also brought accusations of his alleged pro-Communist sympathies. An air of McCarthyism tinged the campaign to remove Bowen, reaching a crescendo when Bowen was accused of simply being too liberal to run a business school. Bowen's book *Social Responsibilities of the Businessman* would be published by the Federal Council of the Churches of Christ in 1953 and was a landmark early volume in what soon became a core topic in the nation's business schools and corporate corridors. By 1964, Bowen was highly regarded for his work at the University of Iowa and gained a reputation as both a scholar and an administrator.[26]

While Bowen kept his nomination quiet during the fall, White House staff prepared a list of the other commission nominees. Yet Johnson refused to make any public announcement prior to the election. His well-honed political antennae were instinctive when it came to timing. He knew that every appointment had both political costs and benefits, and he wished to offend no one before Election Day. Eleven days after his landslide defeat of Barry Goldwater, however, he announced his appointments. In the identical November 14 telegrams to each nominee, Johnson reiterated the balance between the core values of technology and jobs. "We must learn to utilize and expand the opportunities that technology creates," he wrote. "But we must also provide effective means for dealing with the accompanying problems." Since Congress was in election recess, these so-called "recess appointments" would wait for confirmation once the Senate reconvened in January. On January 6, the president sent his nominations to the Senate, where their confirmations did not take much time to approve. The Senate Labor Committee approved the nominations en bloc on January 25, 1965, and the full Senate confirmed them two days later.[27]

Gathered for a formal swearing-in ceremony in the Fish Room of the White House on January 29, the fourteen commissioners were greeted by the new vice president, Hubert Humphrey. Though suffering from a bad cold, the indefatigable former senator, who had offered his own commission legislation just the year before, seemed full of his usual jauntiness. He took the hand of his longtime friend Anna Rosenberg, walking her forward to become the first commissioner he would swear in. Rosenberg had served under FDR when he was governor of New York, and was an assistant secretary of defense in the Truman administration. Republican New York City mayor Fiorello La Guardia would say that "she knew more about labor relations and human relations than any man in the country." She gained a national reputation for her fair-mindedness when she was asked to help mediate the city's 1959 transit strike.[28]

As appointed, the commission would reflect a balanced group consisting of three representatives of the major labor unions and four CEOs of major US corporations, reflecting the labor-management "balance" that the authorizing legislation only vaguely required. In addition to Johnson's friend and labor confidant Walter Reuther, labor's representatives included Joseph A. Beirne, the president of the Communications Workers, whose members included a shrinking core of telephone operators impacted by automation. Beirne had told Congress that "we must be the masters, not the servants of technological innovation so that its unlimited potential benefits will be directed toward the advancement and betterment of all mankind." Albert J. Hayes, the president of the International Machinists, whose members were on the front lines in the automation debate, was the other labor appointee. Hayes had told Clark's subcommittee in December 1963 that automation seemed "a revolutionary force conceivably capable of overturning our social order." If the commission did nothing more than draw public attention to the problem, Hayes stated, it would be a good investment.[29]

Though labor would have only three representatives as compared with industry's four, the appointment of Whitney Young, president of the National Urban League, assured that there would be another strong voice seeking the restraint of unfettered automation. Throughout the many hearings on unemployment, there was frequent reference to the disproportionate impact of automation on the country's African American population. At a conference on the subject in December 1963, Young noted that with some 15 percent of the black workforce still unemployed, automation took its largest toll on African Americans. Young was at the forefront of civil rights legislation, yet he relentlessly made the case that without jobs, blacks would experience "a mouthful of civil rights and an empty dinner table." At the Urban League's closing banquet in September 1962, he noted that "industrialization and automation will form new segregation patterns that are going to challenge integration efforts for years to come." Just as King repeatedly warned, Young also emphasized that the civil rights bill would become an empty symbol unless it translated into tangible economic gains. Equality, he said, must extend to the availability of jobs and economic security for all Americans. As a devoted supporter of Johnson, and one the president later confided in as he became ensnared in the war in Vietnam, Young's perspective on the impact of automation made him not only a logical but also a much-needed liberal-labor appointment to the commission.[30]

The four corporate representatives Johnson chose reflected his strong support for the continuing development of innovative technology. Edwin Land, the scientific genius behind the instantaneous picture and the creation of the

Polaroid camera, was emblematic of the sort of breathtaking new discovery transforming the economy. In his lifetime, Land personally would amass 535 patents. Unfailingly dedicated to technological progress, he was always at the cutting edge of new scientific discovery. Patrick Haggerty was the cofounder and president of Texas Instruments, a company already known for advancing transistor technology. Haggerty had joined Texas Instruments' predecessor company following a stint in the Navy airborne electronics unit. Under his leadership, the company became the largest producer of semiconductor devices in the world. John I. Snyder, chairman of US Industries, pioneered equipment used to help automate industry, including a device to help retrain the human mail sorters that his machines eliminated. "In my opinion," Snyder said, "we have thus far been guilty of a general failure to face the problems of automation, and this failure represents a national moral weakness itself." When he died in April, leaving a vacancy on the commission, the *New York Times* stated that "no American industrialist showed greater sensitivity to the enormous human problems engendered by automation." Finally, Philip Sporn, the retired CEO of the American Gas and Electric Company, was a strong advocate of atomic power. But when he testified at Clark's hearing earlier in 1964, he warned of automation's "potentially disruptive influences on every segment of our energy economy." His unabashed enthusiasm for atomic power did not override his understanding of the human toll that automation was taking in coal states.[31]

Johnson also chose two rising intellectuals. An excerpt from the sociologist Daniel Bell's soon-to-be-published *The Coming of Post-industrial Society* received almost cult status as it circulated widely in draft among academics, policymakers, and public intellectuals. Bell characterized a transformed future economy where business elites were losing power to a new technocracy. According to Bell, the labor market was transitioning from serving manual laborers to clerical workers, technology was transforming life at an unnerving pace, and the old industrial class system was being supplanted by a new order that separated the highly educated from everyone else. Bell would challenge members of the commission with his new vision of the future. So, too, would another budding intellectual force, Robert Solow, back at MIT after a stint on Kennedy's Council of Economic Advisers. The eventual Nobel laureate had already won a prestigious award as the nation's best economist under the age of forty. His 1957 article "Technical Change and the Aggregate Production Function" established a new methodology for measuring the contribution of technology to productivity and thus economic growth, an insight that seemed well suited to the commission's task.[32]

Like Bell and Solow, who were seen as independent voices, the president's appointment of Benjamin Aaron, a longtime labor negotiator, also seemed to be an effort by Johnson to create a strong, impartial center on the commission. Aaron was a highly respected labor mediator, having recently been selected by President Kennedy to help ease tensions during the threatened rail strike in 1963. FDR had named him director of the National War Labor Board, which established workers' wage guidelines while seeking to avoid strikes during World War II. Later, while Aaron was serving as the director of the Institute of Industrial Relations at UCLA, Truman tapped him to serve on the National Wage Stabilization Board, which sought to stabilize wage agreements to avoid inflation. With Aaron, Robert Ryan, and Anna Rosenberg, the Automation Commission was well positioned with members who might help resolve the differences that would inevitably arise between labor and industry representatives.[33]

This stellar panel of "the best and the brightest" seemed exactly what Kennedy originally had in mind when he began talking about an automation commission on the campaign trail in 1960. To assist Bowen with his duties as chairman, Johnson was required to appoint an executive secretary, and the search began. In a December 8, 1964, letter to Bowen, Johnson staff assistant John B. Clinton described the need for a "first rate, technically competent economist with a high degree of integrity, objectivity, public policy sensitivity and strength of character to understand and sort out the problems and pressures surrounding the Commission's area of study." Conspicuously absent among the names being considered was Wirtz's personal favorite, Garth Mangum, who was described by White House staff as "very high on Washington experience." Mangum was then serving as executive director of the President's Committee on Manpower in the Department of Labor, had considerable administrative experience, and most importantly had Wirtz's backing. In addition to his current post in Labor, Mangum had served as Senator Clark's research director throughout the extensive 1963 "Nation's Manpower Revolution" hearings. Born in Delta, Utah, in 1926, and with a PhD in economics from Harvard, he also taught economics at his alma mater, Brigham Young University. Though it would be March 5 before the Senate officially confirmed him, Mangum began effectively acting as executive secretary in late January.[34]

The Work Begins

As they began working together, it became clear that Bowen and Mangum had a good working relationship built on mutual respect and trust. As a

Washington outsider, Bowen needed someone like Mangum, with his Washington insider experience. Mangum had developed a reputation as the consummate staffer in service of his superiors. Shortly after his appointment, Mangum made what he described as the "rugged trek" to Iowa City to meet with Bowen and begin planning for how the commission should proceed. Bowen would not be alone in finding Mangum an indispensable asset to the commission. The technology expert Frank Lynn, asked by the commission to conduct one of its major analyses, said that Mangum "was so hard working that it inspired the rest of us." Lynn observed that as a devout Mormon, "Garth was not the sort of fellow who went out for a beer after a commission meeting. Instead, during long weeks working on the Commission's many simultaneous projects and analyses, he spent his weekends building churches."[35]

Because the commission was scheduled to meet only a few times over the course of the year, Mangum became indispensable in providing substantive continuity between meetings. In addition to culling useful analysis from existing information, he arranged for additional analyses, organized witnesses, developed meeting agendas, and engaged in drafting and editing the final report. When Mangum constructed an outline for the first meeting agenda, Bowen quibbled over emphasis rather than substance. Given the short time frame the panel was given to produce its recommendations, Bowen and Mangum's close working relationship was critical. After their initial meeting, Bowen wrote Mangum expressing what a pleasure it was to have him in Iowa City. "I counted it a real privilege to get acquainted with you," Bowen told his new executive secretary, who would provide the research and writing necessary to keep the commission on task and on schedule.[36]

Guided only by what the authorizing language required of the commission, Bowen and Mangum designed a work plan that entailed two-day meetings every month, with a goal of circulating a draft report by September 1. Mangum and his staff assembled reading materials for the commission members that were mailed a week or so in advance of the meeting. While the commission occasionally heard from live witnesses, the majority of outside input resulted from the commission's request to hear from the public in writing, a request that produced the views of some one thousand separate organizations representing all aspects of society. The commission also contracted for thirty-four separate research studies, though given the shortness of time, Bowen acknowledged that many of these "studies" would not rely on any new research. The Bureau of Labor Statistics and other arms of the Interagency Committee serving the commission conducted some of these analyses. Some were assigned to experts in the private sector, such as Lynn, an emerging expert in the area of technology diffusion. Juanita Kreps, who would go on to become secretary of commerce for President Jimmy Carter, collaborated with her colleague Joseph Spengler on an analysis of "the impact of technological change on the distribution of goods and leisure." The commissioners themselves also contributed analyses. For example, Bell submitted a study on a system of "social accounting" as an emerging means of quantifying technology's impacts on labor, the environment, and communities. Meanwhile Solow addressed the impact of wage differentials on lower-skilled employees. There was no lack of information available, and Mangum capably coordinated and packaged it for commission members to consider in their compressed time together.[37]

Even Johnson offered his own input, though he played what might be considered a customary role for a president consumed by far more immediate priorities, including a growing conflict in Southeast Asia. By creating the commission, both Congress and the president had conveniently "parked" the automation conundrum away from a crowded political stage for at least a year. Figuratively out of sight, the issue was conveniently out of mind. When Johnson's close aide Jack Valenti passed along a request from Wirtz to the president to stop by the commission's first meeting to show interest and "stir them on," Johnson respectfully declined. The president did, however, formally submit to Bowen the report of his Advisory Committee on Labor-Management Policy from January 1962. In doing so, Johnson stressed how Kennedy's advisory committee revealed that employee retraining under the Manpower Development and Training Act would reach only those who had lost jobs, not those who might lose them, thus hinting at what might be an area for the commission's consideration. Moreover, the advisory committee's report had stressed the need for advance notice of factory closings, creative forward planning by firms to avoid layoffs (i.e., early retirements and transfers), generous severance packages, and supplements to unemployment benefits (such as the GAW). The Kennedy panel also agreed that although the impacts of automation could be attenuated by simple attrition, early retirements, and job transfers, there was mounting pressure for more imaginative solutions.[38]

Awash in analysis and information, Bowen and Mangum's challenge was to keep the discussions focused. The legislation required the commission to answer five basic questions. First, it was to examine the pace of technological change, still elusive after so many previous analyses. Second, it was to examine how automation was influencing the qualifications for employment, and whether new skills requirements would impact employment patterns in the future. Here the commission was to pay particular attention not only to specific regions and industries most affected by automation, but also to how the rapid acceleration of technological development was impacting families and communities. Third, reflecting Senator Hart's influence, the commission was to address how new technologies, particularly those developed as a result of federal research, might actually be applied to meet the pressing needs of communities. Fourth, it was to assess where effective technology transfer could yield the greatest benefit, and how public and private funding might be mutually reinforcing in advancing new technologies that would meet both individual and community needs. Finally, Congress instructed the commission not to limit its recommendations to management and labor, but to broaden its sights by proposing how federal, state, and local governments could support the advance of technology that would lead to "continued economic growth and improved well-being of people." Though the commission had a $1 million budget to spend, and liberally farmed out work to experts and consultants, the sheer task of gaining group consensus on such an expansive charter, pulling together meaningful recommendations, and then preparing a full report to Congress in the span of a year seemed enormously challenging. But after four decades of chasing after the unpredictable vagaries of new technologies that were transforming the workplace, the commission seemed to represent the last best hope of realizing the liberal-labor vision of a more planned response to the growing onslaught of computerized operations.[39]

CHAPTER 8

Bold Solutions

*For in your time we have the opportunity to move not
only toward the rich society and the powerful society, but
upward to the Great Society.*

—President Lyndon B. Johnson, remarks at the
University of Michigan, May 22, 1964

On January 4, 1965, as President Johnson headed
toward Capitol Hill to deliver his State of the Union address, he could look
back on a remarkable year of legislative accomplishment. Yet with the
momentum of an overwhelming electoral mandate, he wanted to brand the
reinvigorated liberal-labor agenda in his own words. The previous spring,
in a series of commencement speeches, the president had referred to what
he called the "Great Society." Now, two weeks before his inauguration, he
lumbered triumphantly toward the House dais to outline his vision of a bold
domestic agenda and what he described as "the greatest upward surge of
economic well-being in the history of any nation." Reinforcing his electoral
sweep of forty-four states, he applauded "a unity of interest among our people that is unmatched in the history of freedom," and declared the country
"at the beginning of the road to the Great Society." Two weeks later, in his
inaugural address, he noted that in a nation that had just sent a rocket toward
Mars (*Mariner IV*), "families must not live in hopeless poverty." For the thirty-
sixth president, the line between the impoverished and the unemployed was
indistinguishable, and neither had a place in the Great Society.[1]

The still-bright glow of Kennedy's idealistic legislative agenda, now coupled with Johnson's political acumen, held forth the promise of realizing
greater economic and social justice. The Automation Commission itself was
asked to design policies that might seize the potential of an automated future

while alleviating its corrosive impact on jobs. The commission's expansive charter reflected the energetic vision of the New Frontier, now joined to the goal of achieving Johnson's vision of a Great Society. The attendant hubris of the era would permeate the commission's deliberations and its final report. From the very outset, the commission asserted, "Our problem is to marshal the needed technologies, some of which are known and some not yet known. If we are to clean up the environment, enhance human personality, enrich leisure time, make work humanly creative, and restore our natural resources, we shall need inventiveness in the democratic decision-making process as well as in needed technologies." Commission members were mindful of the fact that they needed to think beyond the boundaries of existing policies as they addressed how automation might help shape the Great Society. For forty years, the solution to the problem of technological unemployment had eluded policymakers. Now the commission was being asked to formulate ways to harness the impacts of automation in just a few months.[2]

Sworn in a week after the inauguration and holding a brief meeting to organize and plot a work plan, the commission convened its first substantive two-day meeting in Washington, DC, on February 18. They met in a capital city itself fully subscribed to computer technology. In 1954, when the Machinery and Allied Products Institute had derisively referred to "automation hysteria," the government was using only ten of the cumbersome, first-generation mainframe computing contraptions, and then only for menial clerical tasks. A decade later, however, the federal government proudly boasted the use of some two thousand computers. Nationally, the number was estimated to have grown from a mere dozen to nearly fifteen thousand in ten years. Indicative of the exponential growth in computerization, some ten thousand additional new computers would be produced during the year the commission deliberated. Companies like IBM and Bunker Ramo were becoming household names. Having already taken over many office functions, an improved generation of computers were steadily advancing onto the factory floor, no longer simply to analyze overall production rates, but more importantly to take over running entire operations, as indeed *Fortune* had envisioned twenty years before. As the *Wall Street Journal* reported, "The computer, through a series of electrical impulses collected by wires from dozens of instruments that measure temperature flow, pressure, and other variables, is taking over responsibility for the whole industrial process."[3]

As with the technological advances in previous decades, job loss inevitably accompanied the growth in computer use. And the losses were not suffered only by factory workers. The loss of clerical jobs nationwide was estimated to be as high as 350,000 annually. In industry, the effects were just as significant. For example, in the appliance industry, while overall productivity had increased over 50 percent in the previous decade, the number of employees declined by the same amount. Even those companies making the very equipment that populated the newly automated plants were losing employees. Employment in the instrument production industry had declined some 15 percent since 1958. In a statement submitted to the commission, Walter Buckingham, an expert on automation at Georgia Tech University, noted that it took two-thirds of a century to develop the electric motor, but only three years from the discovery of the first transistor to the device's widespread commercial availability. Buckingham said that although society seemed able to gradually adjust to the economic effects of automation, it seemed incapable of addressing the destructive impacts that automation was having on communities and personal health. This failure was particularly acute in its effects on African American communities, which were already characterized by unemployment and excessive poverty.[4]

Thus, as the issue of technology and jobs took on new dimensions, commission chairman Howard Bowen faced the unenviable task of having to focus a group of opinionated leaders on a range of challenging mandates. Bowen seemed attuned to the political reality that the commission, regardless of its success in tackling these issues, would inevitably be criticized for comprising a group of unelected experts insulated from public opinion. The chairman thus directed the commission to solicit public comments from the public while engaging thirty-four separate external studies that ultimately were incorporated in appendices to its final recommendations. While the commission would hear testimony from only fifteen witnesses, it received some one thousand public comments and published the most significant statements it received as a separate appendix. The statements represented a diverse range of opinions from unions, corporations, consultants, and academics commenting on the current impacts of technological change and projecting forward ten years, as Congress had requested. Each was asked by Mangum to define not only how to ameliorate the impacts of automation over the next decade, but also how technology might be harnessed to solve unmet community needs.[5]

The Public Debate

With few exceptions, the public responses the commission received reflected the familiar and well-worn patterns that had characterized the technology-employment debate for the past four decades. Since the vast majority of comments came from unions and companies, they exposed the continuing

schism between management and labor and their divergent views on automation. For the most part, industry predictably praised automation's contribution to increased productivity and reassured the commission that the jobs
being created would soon overwhelm those being eliminated. With some
notable exceptions, organized labor continued to raise concerns about the
pace at which automation was occurring and the alarming rate at which it
was taking jobs. Neither side, however, seemed particularly eager to comment on the larger issue of how technology might assist in addressing larger
societal problems, and with the Manpower Development and Training Act
(MDTA) now in place, industry by and large conveniently ignored that issue,
turning instead to its broader concern about the inadequacy of the nation's
education system to meet the needs of an automated economy.[6]

For example, mindful of growing Japanese competition in steel production, the Inland Steel Company urged greater government investment in
education that aligned with future industrial needs and would help US companies maintain their competitive edge. The Iron and Steel Institute reinforced the notion that industry employment was based on customer demand
for US-made steel far more than on automation. Thus, the growing threat
to that demand was foreign competition, particularly from Japan, and not
from new automated steelmaking equipment. At the same time, the steel
producers implicitly acknowledged that their agreement with the United
Steelworkers accounted for automation's impacts by allowing for the transfer of workers to new jobs during down cycles, for early retirement, and for
supplemental benefits in case of layoffs. Other companies, not mentioning
how strenuously they resisted such union demands, nonetheless indicated
having taken similar steps intended to mitigate the impacts of automation.
For example, the Swift Company pointed to the meatpacking industry's policy of allowing the transfer of displaced workers to jobs at newer facilities,
ample separation allowances, and provisions for early pensions.[7]

Generally speaking, industry remained bullish on the benefits of automation. Speaking on behalf of its diverse industrial membership, the National
Association of Manufacturers even went so far as to dismiss the claim that
automation fell hardest on the unskilled. It asserted that since 1963 the
employment of unskilled laborers and semiskilled operators had increased at
a faster rate than general employment. The umbrella business organization
reported that its member companies were becoming more adept at moving employees from obsolete positions to new operations. Such a rosy view
of automation's overall impact, however, was contradicted in statements by
some of the group's own members. For example, FMC Corporation, a diver-
sified conglomerate, was frank in admitting that for its customers in the food

processing industry, its automatic processing equipment would eliminate some fifteen thousand jobs. Though the company noted that it was making canned goods more affordable for consumers, it also candidly acknowledged that "the primary effect of this [employee] displacement will be on the relief, welfare, and unemployment rolls of the communities where the displacement occurs."[8]

Yet such admissions were the exception. Most industry statements predictably sought to tout the benefits and ignore the job loss attributable to automation. In fact, General Electric questioned whether there were any deleterious effects that could be attributed to automation. The company regarded ideas such as "share the work" as completely unnecessary and the result of "faulty political calculations." Instead, GE urged tax policies that encouraged greater "business outlays" in new equipment while recommending improved public education that would enhance worker skills and keep the country competitive in the long term. Indeed, the call for stepping up the quality of US education was a familiar refrain among many industry respondents. GE's rival conglomerate Honeywell underscored the need for greater investment in forward-looking educational opportunities that recognized the country's need to be more technologically competitive. In a prescient observation, the company noted that regardless of a worker's current skill level or job title, the demand for new skills would be changing so fast in the future that it would require continual learning to remain employed.[9]

Those in the expanding service sector of the economy were equally bullish on automation and dismissive of government remedies. Insurers such as Metropolitan Life, Equitable, and Prudential argued that it was the government's own investment in computerization and data processing that was spurring dramatic increases in productivity in an emerging sector of the postwar economy. From punch card sorting, to the replacement of manual calculations, to the full-bore implementation of computerized data processing equipment, the insurers noted that in the postwar era they had lowered costs to policyholders, increased their customer base, and added rather than subtracted employees. Northwest Mutual noted an 8.3 percent increase in employees since 1956. Like others in the new service economy, the insurers were particularly forceful in urging an increased federal role in supporting education. Prudential had its own program for high school dropouts, helping them to support themselves while they trained for jobs. Elsewhere in the service sector, United Airlines observed that as a result of automation, the growing airline industry was predicted to add some fifty thousand new employees by 1970.[10]

Yet of all the sectors of US industry, understandably among the most effusive about automation were those companies in the emerging information economy, increasingly branded by their development and use of computers. Burroughs Corporation, which for years sold hand-operated adding machines, was now transforming office work with new computerized machines. Ray Eppert, who joined the company in 1921 as a shipping clerk, was now its CEO and a friend of Johnson's close aide Jack Valenti, whom he copied when submitting the company's formal statement. Burroughs's statement characterized the new information technologies as the creative vehicle that would spur even more robust economic growth. Noting that automation was a catch-all term, the company suggested that it touched virtually all sectors of the economy, from new automatic mining machinery to computerized inventory controls. In the aggregate, the company calculated that these diverse technologies created 11 percent more jobs among the Fortune 500 between 1956 and 1963, and it predicted that rate would increase. To the extent that technologies eliminated jobs, Burroughs argued that the sophisticated new computing capabilities were capable of matching the newly unemployed with new opportunities. Thus, rather than shrink from the productivity increases afforded by new technology, Burroughs bullishly endorsed tax revisions that encouraged more and faster modernization.[11]

It was a perspective shared by Xerox, another company helping define an emerging information industry. Xerox, the company whose very name would soon become a verb, was among those advancing information technologies that were combining to make technological change itself more predictable and controllable. In its statement, the company said it envisioned a total realignment of the economy over the next decade, both in the way goods were produced and in the employee skills necessary to support them. It predicted that soon all production would be done by a small fraction of the total population. The rest would be a part of a new economy engaged in distribution, support activities, personnel services, and education. It was precisely the sort of futuristic vision that commission member Daniel Bell would refer to later as the formative components of a "post-industrial" world. In Xerox's view, societal failures associated with the transition to a new economic alignment did not rest at the feet of technology. Rather, they were due to failures of the political, managerial, and social systems. Going well beyond other company statements, Xerox suggested that the entire retraining imperative was illusory. Suggesting a fault line that was emerging as computer technology accelerated the rate of change and rendered some employee skills obsolete, the company seemed pessimistic that any job training program, as recently authorized in the MDTA, would have any real-world impact. Rather, the cold reality was that "at the point at which a worker suffers technological unemployment, it is already too late to retrain him." Such an analysis from a company on the cutting edge of technology suggested the pressure on the commission to think beyond the present and design policies that anticipated this future rather than catching up to a vanishing industrial past.[12]

For their part, the unions were every bit as resolute and specific in reiterating their deep reservations about the onslaught of automation. The United Mine Workers said its members ranked it as the foremost problem of the time. The Office Employees International Union submitted that "routine clerical occupations were being eliminated at an alarming rate." And the Longshoremen reported that eight years after their dramatic work stoppage in 1958, they still could not report any progress regarding dramatic declines in their jobs due to automated cargo handling. The United Federation of Postal Clerks, not beguiled by statements of reassurance from the postmaster general, feared that the adoption of optical scanning devices and mechanized letter-sorting capabilities would lead to productivity increases that would outpace the rate at which mail volume was increasing. Accepting the inevitability of continued declining industrial employment, the International Union of Mine, Mill and Smelter Workers went so far as to endorse the agenda of the Triple Revolution, suggesting the need for massive public investment in housing, environmental, and public works projects, and the more equal distribution of wealth, particularly to black Americans.[13]

Yet just as there were outliers in industry, not all unions shared the pervasive worker anxieties over automation. The Airline Pilots Association actually welcomed automation. It saw devices such as the "automatic pilot" as helping the industry fly through inclement weather, thus avoiding delays and cancelled flights. Local government employees likewise did not feel as threatened. The American Federation of State, County and Municipal Employees stated emphatically that although automation might destroy jobs in some industries, it was not the case for those in local government. Though a canvass of members revealed that over 63.7 percent of local government agencies installed "electronic data processing equipment," of those using the new technology, 44.2 percent showed an increase in the number of employees (38.5 percent showed no change, and only 17.3 percent claimed an actual decrease in employment).[14]

Yet while there was some division among various union shops over technology's impacts on jobs, the pace at which automation was occurring seemed a universally shared labor concern. The Retail Clerks Association noted how new data-input devices and the use of computers for logistics and for inventory control were transforming the retail space. Though lacking any hard statistical evidence of job loss, the clerks' union cited the BLS's analysis that retail productivity had increased 3 percent annually between 1947 and 1960. This increase in output efficiency was accompanied by a 17 percent rise in retail man-hours. There was no doubt the trend would continue, and the union asked that the commission seriously consider how to modulate the pace of the transformation. The Pulp and Paper Union voiced similar concerns regarding the relentless pace of automation. It witnessed production jobs increasing by only 10 percent even though the industry's productivity increased some 40 percent in the decade ending in 1962. As a result, the union recommended that industry extend the period that would elapse between the decision to invest in new labor-saving equipment and its actual installation, in order to minimize job loss. To achieve a healthy stasis between technological advancement and employment, the pulp and paper workers also urged that management be more forthcoming early on about its modernization plans and its planned pace of automation.[15]

The Pace of Technology

The sheer number and substance of the statements the commission received achieved Bowen's larger purpose of providing at least the appearance that the commission was not a sequestered cadre of elitists. In addition, in seeking to go beyond the well-established boundaries of the automation debate, the commission was informed by a number of separate analyses it commissioned. Among the most important, as Daniel Bell later noted, was the one completed by Frank Lynn. Bell would continue to draw from Lynn's analysis in describing the contours of "post-industrialism" in his much-heralded *The Coming of Post-industrial Society*, published in 1973. Lynn revealed how the elapsed time between the discovery of new technological innovations and their commercialization had decreased dramatically, from over thirty years to nine years, since the early part of the century. Indeed, Lynn provided a new methodological approach to help anticipate the rate at which new technologies might eliminate jobs—the seemingly insoluble riddle that had haunted the technological unemployment debate since Ewan Clague first undertook just such an analysis for Hoover's Presidential Advisory Committee on Unemployment.[16]

Returning from the Korean War, Lynn entered Loyola University on the GI Bill, then transferred to Northwestern, enrolling in the MBA program in 1956. There he did his thesis on the patterns of demand for air conditioning based on an analysis of national temperature trends aligned with industry sales data. The paper drew interest within the industry and from the Illinois

Institute of Technology, which in turn hired him in a new consulting group that quickly evolved into a highly profitable venture known as INTEC. Lynn's work on technological diffusion gained INTEC a reputation equal to those engaged in similar work at the Stanford Research Institute and A. D. Little. While the commission heard from other industry consultants on the pace of technological diffusion for specific industries, it was Lynn's analysis that revealed how the acceleration of technological development, accompanied by increased productivity, was clearly responsible for the structural unemployment plaguing the economy.[17]

Lynn analyzed the trajectory of twenty technological innovations in fields as diverse as electronics, vitamins, integrated circuits, and nuclear power generation. He traced each technology's path from the laboratory to trial application to commercialization. For example, in the electronics industry, he asked how much time it had taken for vacuum tubes, semiconductors, and integrated circuits to move from breakthrough laboratory innovations to products that transformed customer choices. In every case, Lynn described the relative importance of government investment in the development phase and the federal government's role in accelerating the pace of technological diffusion. The exponential increase in federal expenditure for research and development after World War II was playing a major role in accelerated technology adoption, one Lynn confidently predicted would continue as long as government maintained its formidable presence in support of R&D.[18]

Lynn's straightforward analysis of the rate at which new technologies were being adopted provided yet another iteration of the long-standing challenge that was at the center of the technology-versus-jobs debate. It was precisely the sort of analysis the liberal-labor coalition urged in its quest to better anticipate and plan for economic downturns. As Bell later observed, in the microeconomic sense, knowledge of how technology is progressing is critical for any enterprise's survival. In the macroeconomic sense, it was the sort of information long clamored for by those who believed the economy needed to institutionalize economic planning. If only the federal government possessed such a planning capability, it might better forecast and respond to spikes in unemployment. Indeed, as early as 1928, one of Senator Wagner's original bills was aimed at the orderly planning and provision for taking steps to ameliorate future downturns. And after the economic collapse of the Depression, creating a federal budgeting mechanism that might better forecast the future was at the core of Senator Murray's 1945 full employment bill. Planning in order to mitigate capitalism's almost predictable cyclical declines became a consistent mantra of organized labor, fueled by Reuther's continued pleas for forward thinking. With Lynn's analysis, the commission was again presented with an opportunity to reinvigorate the idea. In his laborious but insightful analysis of the impacts of technology on employment across industry, Lynn was offering an analytical tool that might become part of a larger federal government role in economic planning to avoid the worst effects of automation.[19]

The Realities of Prediction

If Bell was particularly impressed with Lynn's far-reaching analysis, it was likely because of his own knack for seeing the unseen. Bell was skeptical of the worst predictions attributed to automation, yet at the same time he was convinced it would not only transform the composition of the workforce but also redefine the US economy. A decade earlier, in *Work and Its Discontents*, Bell argued that automation would fundamentally change the nature of the workforce, creating what he called a new white-collar "salariat," as contrasted with the blue collar proletariat. At the same time, he took issue with Wiener's dismal predictions about a reality of "unattended factories turning out mountains of goods which a jobless population will be unable to buy." Bell thought the more realistic scenario was that many workers, in particular the older population that could not be retrained, would be permanently idled. He predicted some communities would be hollowed out. These severe consequences, Bell wrote, could be addressed through flexible and adaptive fiscal and tax policies. The larger issue stemming from automation, as he presciently foresaw, was that companies would increasingly be less worried about their proximity to a large workforce and therefore much more flexible as to where they located their operations. He further suggested that in factories that were basically running themselves, an employee's worth would no longer be measured in terms of output or hours, but in how the employee added value. Now beginning work on what would become his classic *The Coming of Post-industrial Society*, Bell quickly recognized the value of Lynn's forward-looking analysis. Though the commission asked others to examine the pace of technological diffusion, it was Lynn's work, as Bell later recognized, that garnered the commission's attention in its final report.[20]

In addition to Bell's prescience, the commission benefited from the future Nobel laureate Robert Solow. The MIT economist, having just returned from a stint on Kennedy's National Council of Economic Advisers (NCEA), was particularly attuned to how tax and fiscal policies needed to adjust to the new realities of an automated society. Solow argued that innovation needed to be measured, and the measurements incorporated as a basis for predictions of future economic growth. Innovation was no longer a matter for soft conjecture, but rather a hard reality that could be measured. A neoclassical economist in the Keynesian mode, Solow was a believer in the importance of full employment. But he also believed that attaining such a desirable goal required more than marginal adjustments to capital and labor supply. In what became known as the "Solow Residual," he argued that technological advancement and the resulting changes in productivity were essential components of any economic growth analysis. After presenting his breakthrough adjustments to the assumptions underpinning conventional economic growth models, Solow concluded that in the period from 1909 to 1949, gross output per man-hour had doubled, and a whopping 87.5 percent of that increase he attributed to technological change. Moreover, most of the dramatic increase in worker output occurred during the last two decades of the period. Solow's conclusions and his subsequent work on the importance of innovation and technology for economic growth rates (for which he eventually won the Nobel Prize) underscored why the commission considered Lynn's contribution so important.[21]

Lynn showed how it might be possible to look forward from what was known about existing technologies and quantify the pace of technological diffusion across the economy. Yet he also cautioned that the technological future, particularly what Lynn referred to as the "spillover effects" that come from innovation, remained largely indecipherable. Lynn's analysis focused only on known, existing, and already commercialized technologies in specific industries. In addition to unknown new discoveries being made away from sight, even in the moment, the ancillary effects that any one technology might have on other industries, processes, and markets were less clear. For example, it was one thing to predict job losses from the introduction of an oxygen-based steel-producing process, but it was another to determine the impact of such new innovations in steelmaking on the energy-producing industries that fueled the conventional steelmaking process. Even more fundamentally, it remained impossible to see such a new process emerging in the first place. Lynn, however, was less concerned about surprises. He tried to reassure the commission that "although the rate of development and diffusion is increasing, the lapsed time from basic technical discovery to the point where significant social and economic problems become evident is still relatively long." Thus, once a technology emerged from the darkness of the discovery period, Lynn concluded, there was still time to allow for an early-enough warning to mitigate the deleterious employment and community impacts.[22]

Others from outside the commission were not quite so sanguine. On the verge of winning the Nobel Prize in Economics, Paul Samuelson believed that predictions were largely a function of understanding the potential of technology in the moment. It was simply impossible to see what some inventor was doing in some remote laboratory that might transform the future of the economy. For example, in the late 1950s, a new company called Texas Instruments had successfully integrated multiple transistors into silicon through the use of micro-wires. Another company, Fairchild, then used an entirely new process to produce a silicon wafer that at first combined hundreds, then thousands, and ultimately millions of microscopically small transistors. Similarly, in the early 1960s, Larry Roberts decided to pursue the vision of the Department of Defense employee J. C. R. Licklider to realize how a network of computers might be organized to share information in what would soon become known as ARPANET, ultimately to become the internet. Even as the commission met, a German physicist, Manfred Börner, at Telefunken Research Labs in Ulm, made his first patent application for a fiber-optical transmission system. Just as Wiener had been unable to predict how the applications of his own language of cybernetics would enable the automated factory, no one could foresee how any of these four unrelated discoveries would over the next twenty years collectively help to launch a truly postindustrial economy, where the sheer speed of the movement of information, its storage, and its instantaneous availability would become more valued than the physical products of the industrial era.[23]

Shifting Sands

As the commission members continued to deliberate for two days each month during 1965, the sands of political reality steadily shifted beneath their feet. Even as Lyndon Johnson was appointing the commission in December 1964, manufacturing jobs topped sixteen million for the first time. A month before Bowen was giving his midyear report to the president, Labor Secretary Wirtz reported to Johnson about the improving economic news. The unemployment rate, which was 6.9 percent when Kennedy was inaugurated, now stood at 4.6 percent, its lowest since 1957. In fact, as Wirtz summarized for the president, the sustained recovery was marked by fifty-one consecutive months (February 1961–May 1965) of economic expansion, with the number of jobs increasing by 5.3 million and the number of jobless Americans declining by 1.4 million. The numbers also revealed that "non-white" Americans experienced a 40 percent decline in the jobless rate during this same four-year period. Wirtz indicated that not only was the number of unemployed diminishing, but since Kennedy's inauguration the number of those with jobs increased by almost 4 million. Wirtz proudly reported that every worker group that the department analyzed showed an increase, except for teen-age workers. The lack of employment opportunities for recent high school graduates would become a concern of the commission.[24]

Meanwhile, Wilbur Cohen, who was the acting secretary at the Depart-ment of Health, Education, and Welfare, was also reporting the adminis-tration's success in implementing its new job training law. He claimed that twelve thousand adults in nine different states had already become self-supporting because of the act. When their families were included in the figures, it resulted in some fifty-eight thousand fewer people dependent on government support. He promised that training projects now up and run-ning were targeting another forty-three thousand unemployed in forty-four states, and he asserted that the new program was gaining momentum and seemed to be producing positive results. The program, Cohen told Johnson, was particularly beneficial in high-unemployment states like West Virginia and Illinois, whose Senators Randolph and Douglas were both cosponsors of the original legislation.[25]

With the economy rising and unemployment falling, Bowen sensed that the work of the commission was becoming less relevant to a White House increasingly turning its attention to its broader War on Poverty and the real war in Vietnam. Testing the waters, Bowen wrote to the commis-sion's White House staff liaison, John Macy, on September 27, reminding the White House that the deadline for the commission to issue a report was January 1, 1966, and that the commission would then cease to exist thirty days after the release of its final report. Bowen inquired as to what plans the administration had for release of the report, and with his trademark diplo-matic finesse, he concluded by reminding Macy that fourteen particularly busy individuals had dedicated considerable time to the commission's work. At the same time, he said, the members clearly desired "the release of the report to mesh with the program of government." After circulating Bowen's letter, Macy quickly responded that the White House would take ownership for any public release and reassured the chairman how critical the commis-sion's work was.[26]

It was not as if public suspense were mounting as the commission neared its end-of-year deadline. By late November, a few press accounts did appear to suggest that the commission was nearing its end and hint at what the forthcoming report might recommend. But even leaked drafts did not command much press attention. The generally sparse press cov-erage tended to highlight the internal disputes emerging among commis-sion members over how the final report would characterize the overall impact of automation on jobs. Only the New York Times, as Bell later noted, had followed the commission's work from the outset, and it previewed the impending report with a front-page story detailing its preliminary recommendations as compiled from various sources. In its Christmas Day edition, the *Times* suggested that the commission was prepared to recommend some "striking ideas on what can be done to reduce the future toll in human casualties" from the onslaught of automation. Indeed, the commission would do just that. But when the report finally issued officially in early February 1966, the boldness of many of its recommendations would continue to receive scant press coverage, or any congressional attention whatsoever. The commission that was created out of Congress's own self-acknowledged inability to get ahead of the technology curve delivered its recommendations as employment was rising, and concerns over Vietnam were taking on new political relevance. As the final report itself acknowledged, the administration's aggressive fiscal policies had altered the political context—or, as it said succinctly, "Conditions have changed, and public concerns have been modified." In short, as they concluded their deliberations, the commission members seemed well aware that there was little public appetite for their findings, let alone their bold and visionary recommendations.[27]

Managing Differences

As with virtually any diverse group, reaching a final consensus on both findings and recommendations required skillful leadership, particularly as the deliberative process wound to a close. The commission was being asked to comment on drafts prepared by Mangum and his staff, with the able input of Solow, Bell, and Benjamin Aaron. Aaron's capabilities in managing labor-management disputes were put to the test in the closing weeks, as the conflicting views over the impact of technological advancement threatened to derail any hope for a consensus report. The union representatives—Hayes, Reuther, and Beirne—thought the draft report soft-pedaled the impact of technological acceleration and the ability of the economy to absorb it, thus minimizing any concerns over long-term job impacts. When their concerns were leaked to the press at the December AFL-CIO convention, the press finally had just the conflict it needed.

Though the three labor leaders were not possessed by "incoherent anxiety," as one account described it, they were genuinely irked that automation was characterized as not posing the largest threat to future job growth. They stood firm in recommending a thirty-five-hour workweek, which required Bowen and Mangum to attempt to artfully reach a compromise with business representatives. At the last scheduled commission meeting on January 9,

1966, the commission's four business leaders (Watson, Land, Sporn, and Haggerty), sensing they were about to come under pressure, pointed out many of the concessions they thought they had already made. Seeing little agreement between the two sides on the tone the final report would adopt, and faced with the prospect of a report that its liberal-labor members would not sign, Bowen offered his solution. He recruited some of the business representatives to express their desire to make dissenting comments at the end of those sections where they took issue. Then he asked labor if it would join in that process throughout the text. These notes, sprinkled throughout the eventual 115-page final report, allowed both labor and business a platform to air their views while still giving the commission an appearance of overall unity.[28]

In fact, the reported divisions were likely overemphasized by reporters in an effort to fuel more interesting press accounts. As it reads, the commission report does indicate broad consensus on many issues, though as with any compromise document it often skillfully genuflects to conflicting positions. For example, on the contentious issue of whether automation was causing unacceptable job losses, the sheer thoroughness with which the report addressed the subject, as well as many others related to it, provided ample ammunition for both automation's defenders and its detractors. On the one hand, based on Lynn's analysis, the report asserted that the pace of technological advancement would remain steady without any dramatic step-change increase over the next decade. Thus, business interests could safely suggest that automation would not lead to sudden spikes in unemployment and allow for the retraining of those who did lose work. At the same time, the commission acknowledged that there would have to be an unprecedented rate of economic growth to achieve sustained levels of employment, thus allowing labor to continue to raise its concerns regarding the manageability of technology's advancement. Such consistent equivocation, however, did result in the three labor representatives continuing to critique the final report's lack of an urgent tone, though they seemed less concerned about specific recommendations or any substantive deficiencies.[29]

Fulfilling the Mandate

Bowen and Mangum created an atmosphere that allowed a room filled with not only intelligence and expertise but also opinion and ego to argue extensively within the expansive boundaries Congress set for the commission, while staying focused on its legislative charter. Congress asked the commission to identify and assess the role and pace of technological change, its impacts on employment patterns, how technology might be utilized and transferred from government and private research to address unmet societal needs, and how the benefits of technology could be more evenly distributed across society. In response, and at the outset of its final report, the commission acknowledged that in only six of the twelve years since the end of the Korean War was the rate of economic growth significant enough to exceed the rate of productivity increase and thus result in growth in the labor force. Thus, the commission served notice from the outset that its findings were not aimed at slowing the growth of productivity or slowing innovation. Left to its own devices, however, the commission anticipated that over the next decade, the economy would have to achieve a sustained 4 percent growth rate in order to maintain full employment. Since it was unlikely that would be achieved, the commission therefore called for aggressive fiscal and monetary policies. It specifically acknowledged that the administration's tax cuts were already helping reduce unemployment, but in response to the labor members on the panel, it called for even more aggressive measures to spur government spending and ease monetary restrictions.[30]

Indeed, as bold as many of its recommendations were, the commission did not shy away from dusting off old ideas such as reliance on fiscal policy, or even public works. In examining automation's impact on the future composition of the workforce, the commission did not forecast a decline in unskilled labor. At the same time, however, it recognized that automation was eroding skilled labor positions. The commission thus called for new educational and training initiatives to provide the younger generations with the essential skills necessary to secure and retain jobs in an automated economy. It was concerned about the challenges presented in preparing young people, particularly African Americans, for the workforce. "Our society must do a better job than it has in the past of assuring that the burdens of changes beneficial to society as a whole are not disproportionately borne by some individuals," the report stated. Thus, just as in the recent past, when old-guard liberal-labor loyalists such as La Guardia and Clark argued for public works projects, the commission still saw it as essential that Congress make certain that the "hard core" unemployed had work even if government was the employer of last resort. The commission also returned to Senator Wagner's proposal to provide the unemployed with the wherewithal to move to areas where there were jobs, a challenge now made more feasible by the adoption of computerized job-matching capabilities. Similarly, drawing from Senators Kennedy and Douglas's ideas for bolstering depressed areas, the commission recommended going beyond the recently enacted depressed areas legislation with infusions of federally subsidized venture capital and targeted infrastructure improvements for these regions.[31]

Not only was the commission willing to recast previous policy solutions, but it also spared no sector of the economy in its broad-ranging recommendations. For example, it noted that among the hardest-hit populations were farmworkers, whose number, as a result of new technologies, had decreased 43 percent since 1947. As in urban areas, the decline in rural employment was falling hardest on African Americans. Consistent with LBJ's emphasis on civil rights, and with the intent of the Civil Rights Act of 1964, the commission was adamant regarding the need to address the educational gap between white and black Americans. In the commission's unanimous view, the disparities in educational attainment levels were unacceptable. While acknowledging that job retraining initiatives seemed to be going well under the new MDTA, the commission recommended an aggressive new program of "compensatory" education for those from disadvantaged communities, including in the rural United States. This recommendation aligned well with the White House's War on Poverty initiatives.[32]

But the commission would go well beyond even this expansion of the MDTA to recommend "fourteen years of free public education, elimination of financial obstacles to higher education, lifetime opportunities for education, training and retraining, and special attention to the handicaps of adults with deficient education." Wirtz, who was cochairing the administration's advisory panel to the commission, had previously recommended two additional years of free education beyond high school, and both the National Education Association (the primary teachers' union) and the American Association of School Administrators had promoted the idea. Again, Mangum's knowledge of already completed government analyses he helped prepare while at the Department of Labor helped smooth out any resistance to this bold recommendation. Emphasizing the coming "skills gap" over the next decade, Mangum drew the commission's attention to the projected high levels of youth and African American unemployment. From that point of departure, the commission members agreed that it was imperative that upper-level education be made freely available to all.[33]

But extending free public education through grade fourteen was not the only far-reaching recommendation the commission made. Convinced that rising productivity would continue to make possible a wider social safety net for those impacted by technology, the commission recommended a basic income for all Americans. The Ad Hoc Committee on the Triple Revolution had adopted the controversial measure as its centerpiece, and though it came in for its share of public ridicule, such criticism did not deter the commission from recommending it to help make up for the continuing inadequacy of state unemployment compensation systems. At the urging of President

Johnson, the commission had considered enhancing state requirements to extend unemployment compensation benefits over a longer period and to add a provision for compensating the long-term unemployed. But the Commission observed that the overall welfare system was beset with its own set of disbursement issues. Under the existing means-testing provisions, a family could lose 100 percent of every dollar it earned, thus providing little to no incentive for work. Here the commission took issue with the new additions to the Great Society's welfare programs, such as Aid to Dependent Children, as too limited in their scope of coverage and the benefits they provided. As an alternative to the cumbersome and inadequate system, the commission boldly recommended that serious consideration be given to what it termed a "minimum income allowance." The commission emphasized the breadth of ideological support for a basic income, which extended from the liberal economist James Tobin, a member of Kennedy's NCEA, all the way to the conservative economist Milton Friedman, who would eventually serve on President Ronald Reagan's staff. The commission firmly recommended "that economic security be guaranteed by a floor under family income." This meant not only wage-related benefits such as adequate health care, but more importantly "a broader system of income maintenance" that could take the form of either a minimum income per individual or a negative income tax. The commission was quick to point out its recognition that many issues of distributive equity would emerge in attempting to find the right approach to a basic income. Nonetheless, all members were in full agreement that the time had come to commence the debate over providing a basic income for all. If there was no policy or planning panacea that could anticipate the next technological paradigm shift, then it seemed time to become reconciled to structural unemployment and adjust public policy accordingly.[34]

Though the recommendation to extend free public education and to guarantee an income would garner most of what little press attention the final report received, the commission was equally bold in addressing how technology might be directed to meet broader societal needs, as Senator Hart had recommended in his own commission legislation in 1963. In the chapter entitled "Technology and Unmet Human and Community Needs," the commission's report spoke to the era's abiding faith in technology's redemptive force. Characteristic of the pervasive technological hubris of the era, the commission unequivocally stated that "technology has the potential, whose beginnings we already see, to realize a persistent human vision: to enlarge the capacities of man and to extend his control over the environment." The members unanimously agreed that the problems confronting society were amenable to a technological solution, "if we concentrate enough money and manpower on them." They recommended that the federal government utilize new computer technologies to provide diagnostic and patient care to larger segments of the population and to train a new corps of health care workers who might help to do just that. Four years before Earth Day and the creation of an Environmental Protection Agency, the commission decried the foul air and water polluting many of the nation's urban centers and declared that the very technological knowhow that helped create the problem now needed to be part of the solution. Particularly with postwar suburban flight at its apogee, the report emphasized the responsibility of the construction and housing industry to tackle the increasing problem of urban blight, estimating that 8.5 million US households were in substandard housing. Though Johnson's Housing and Urban Development Act had passed in September, soon followed by the creation of the Department of Housing and Urban Development, the commission was focused on how technology might help these latest Great Society initiatives to accomplish their goals. The commission noted that technology was making less expensive and more efficient building materials available to assist with the nation's urban revitalization, and it urged the federal government to leverage its new housing budget to change antiquated building codes. "The creation of mass production housing and the undertaking of large-scale urban reconstruction will create a new industry and many new jobs," the commission predicted.[35]

The Chase Continues

Throughout the nine-chapter, 115-page report, the commission recast old liberal-labor coalition recommendations, offered bold new solutions that seemed beyond what was politically practical, and outlined the contours of how technology might play a central role in creating a Great Society. Prior to reaching a final agreement on the release of the recommendations, however, Reuther and his union colleagues began publicly whispering their strong disappointment that nowhere in the final report was there any recommendation for the establishment of federal economic forecasting and planning capability. Moreover, adding salt to labor's wounds, the commission was preparing to instead suggest the creation of a permanent national goals panel made up of "distinguished private citizens representing diverse interests and constituencies and devoted to a continuing discussion of national goals." The liberal-labor coalition's long-standing belief that the government should more accurately forecast economic downturns in order to ameliorate future job losses was absent from the draft. Instead, it recommended creating a panel akin to the National Goals Commission that Eisenhower established during the last year of his second term and whose recommendations were largely ignored following the 1960 election.[36]

Reuther, Hayes, and Beirne, now joined by Whitney Young and Daniel Bell, thus took strenuous issue with the final draft, insisting that a "democratic" national planning function, not another goals commission, was critical given the pace of technological advancement. This liberal-labor cohort within the commission was aware that in the ongoing Cold War context, the very appearance of support for an enhanced forecasting and planning capability had to be carefully couched so as not to suggest the sort of "centralized" planning function that now characterized countries in the Communist Bloc. Thus, the commission's liberal-labor members emphasized that the planning process would seek broad consensus among diverse constituencies and voluntary agreements on devising and prioritizing future actions to ensure sustained high levels of employment. Assuredly, in the Cold War context of the period, it was not to be an elitist planning body that dictated Soviet-like five-year plans. After much persuasion and cajoling by Bowen and others on the panel, the five eventually were persuaded to summarize their concerns in a strongly worded comment that reflected their frustration:

> It is our firm conviction that some form of democratic national economic planning is essential to the United States to assure prompt meeting of our most urgent national needs in both the public and private sectors. The blind forces of the market are no longer adequate to cope with the complex problems of modern society. . . . What we are able to foresee we should be able to deal with rationally. Planning provides the mechanisms for rational action to make the most effective use of our resources both to solve problems and to make the fullest use of opportunities.

The stark reality, however, was that the politically insulated commission seemed no more equipped to add an economic planning mechanism than the 79th Congress was when it stripped such a provision from Patman's original Full Employment Act. For all its thorough analysis and many controversial recommendations, the commission simply could not agree on the liberal-labor coalition's core policy goal.[37]

As the new year turned, the administration announced that the GNP over the previous year had grown at an astonishing rate of 7.5 percent. Many leading business figures were predicting an even more robust year ahead. Meanwhile, the president was busy asking Congress for a substantial increase in the defense budget in support of the rapidly escalating military engagement in Southeast Asia. Johnson announced that the United States was committed to staying in South Vietnam as long as necessary to preserve freedom. In the growing and now increasingly wartime economy, unemployment was holding steady at just over 4 percent, thus meeting the government's definition of full employment. Public concerns regarding structural unemployment having faded, the White House seemed to have to be reminded why the commission existed in the first place.[38]

Two weeks before the final commission meeting, Mangum called White House staff to again alert them that the final draft of the commission's report would soon be ready. He asked who should get copies of the final draft and, more importantly, how the White House wanted to release the report. Perhaps most critically, given the press attention to the controversies raised by the draft, Mangum reassured the White House that these were settled, and that the final report would reflect a means for addressing dissent without having a separate minority report. Then, perhaps sensing the diminished enthusiasm at the White House, he suggested that perhaps the commission should proceed with its own announcement to save the White House the bother. The potential of a presidential commission suddenly firing off its own press releases sufficiently alarmed White House communications director Bill Moyers that he recommended that the commission hold its final meeting at the White House, where the president could formally receive the report and make comments on its significance. Valenti, in turn, asked the president if he would be willing to receive the commissioners at a White House ceremony, and Johnson said yes. Thus, commission members were hastily told to keep three days open on their calendars in late January for an Oval Office presentation of the report. Then, on even shorter notice, they were summoned to Washington for a ceremony to be held on Saturday afternoon, January 29, an event quickly moved to Monday, January 31, to accommodate the president's fluctuating schedule. But the congratulatory moment was canceled altogether when a brutal winter storm gathered force on the East Coast. Bowen would make it no further than Pittsburgh, while other commission members found their flights to Washington canceled. Returning to Iowa City, Bowen dutifully wrote to the president expressing his regrets that "the big snow" prevented the commission from delivering its report to him in person. Bowen concluded his brief letter by perfunctorily stating his appreciation for having served as chairman.[39]

The report itself was subsequently released by the White House without ceremony on February 3 at the end of Bill Moyers's weekly press conference. The scant press release simply stated that the report had been submitted, named the members of commission, and briefly traced its history on a single page of White House press stationery. The president also dutifully sent a brief letter to each member of the commission thanking them for making a "thorough examination of the broad and complex issues of our economic life and their implications for the future of our society." Bell later wrote that in spite of the snow, he thought the White House's reaction unfortunate. "The commission had made truly strenuous efforts to reach consensus," he noted. Though there was some dissent, in the end he believed that organized labor in particular should be pleased with the final recommendations. Nonetheless, the report seemed to be figuratively buried in the deep snows that had hit Washington. Perhaps its brief political moment was gone. Or perhaps it simply had not arrived.[40]

On Capitol Hill, the reaction to the commission's report was similarly underwhelming. One of the earliest and most enthusiastic sponsors of legislation to establish a commission, Senator Javits, simply asked that a newspaper account summarizing it be included in the *Congressional Record*. Though he indicated he would have more to say about the commission later, he never did. Nor did other members instrumental in the authorizing legislation for the commission have much comment. Senator Clark issued a similarly perfunctory set of remarks, thanking the commission and indicating that his Labor Committee would review the recommendations requiring legislation. Yet there would be no formal hearing. Others were less charitable. Congressman Joe Waggonner, a conservative Louisiana Democrat, took to the floor to chide the recommendations as "deep dish pie in the sky." Critiquing as "socialist" the commission's recommendations for free education through grade fourteen and for providing a basic income, Waggonner remarked, "If it were not for the fact that [the commission] has actually placed this proposal on the President's desk, the entire matter would be so asinine as not to be worthy of our attention."[41]

The release of the report received very limited press, and what few accounts there were ranged from indifferently informative to actively critical. The *Los Angeles Times* suggested that "were it not for the prestige of the Commission members, its recommendations might be ignored because of their radical nature." The *New York Times* said that the panel's final report was strongly affected by the improving economy. Unemployment was projected to be 3.5 percent by year's end. Both the *Times* and the *Washington Post* did give the guaranteed income program particular prominence in their stories. And two of the nation's leading columnists, the liberal Roscoe Drummond of the *Christian Science Monitor* and the conservative Walter Trahan of the *Chicago Tribune*, were impressed by the consensus reached by such a diverse group of business and labor leaders. But such stories were the exception.

With this sort of underwhelming press attention, the commission's work quickly faded from any public view.[42]

Later that spring, however, during an April 21, 1966, lecture at Kent State University, the *New York Times* reporter A. H. Raskin praised the commission's work as "the Magna Carta for human progress in an age of technological liberation." As early as 1955, in an article describing the increasingly automated Ford River Rouge plant, Raskin had noted how the drudgery of work had vanished. Now, over a decade later, he lauded the commission for taking on the challenges of the structural unemployment that still plagued many regions of the country. He noted the commission's insight that the existing system of welfare was creating a protected class of skilled workers while simultaneously exacerbating the plight of young laborers and African American workers unable to find work in the increasingly automated economy. Raskin supported the commission's advocacy of a basic income for those who simply could not find meaningful, well-paying jobs. He told his Kent State audience that the commission's work was a blueprint for how public policy needed to adapt to the automated future. The report, Raskin told the students, was a call for both less anxiety over automation and, at the same time, less complacency about unemployment. Yet Raskin seemed a lone voice in seeing any significance in the commission's findings. As the midterm elections of 1966 approached and the legislative momentum toward Johnson's Great Society stalled, the bold recommendations of the Automation Commission quickly became a distant afterthought, placeholders perhaps someday to be revisited with the next wave of technology. And yet, with its comprehensive and far-reaching set of recommendations, the commission's report finally met, if not exceeded, the pleas of those who long demanded that policy be as innovative as technology. So cutting-edge were many of its recommendations that they now seemed politically too much too soon in the context of a growing economy and rising public concern over the war that was fueling it. It was an ironic conclusion to four decades of liberal-labor reforms, particularly for those indefatigable reformers who chased after automation and the human casualties it left in its wake.[43]

Epilogue
Back to the Future

We must guarantee an equal opportunity to work at fair
wages to every person in our Nation who is willing and able
to work.

—Senator Hubert H. Humphrey, August 22, 1974

The Automation Commission delivered its far-reaching recommendations nearly four decades after Senator Robert Wagner's maiden floor speech in March 1928, in which he outlined modest legislative proposals to count the unemployed, notify them of job opportunities, and forecast future economic downturns. During the Great Depression, Wagner soon stood at the center of a liberal-labor coalition that emerged full blown to expand labor's right to bargain for its share of productivity gains, create a federal role to bolster state unemployment compensation funds, legalize a minimum wage, and limit employee work hours. Though many of these reforms were passed in response to the unprecedented unemployment crisis prior to the war, the champions of the US working class continued their crusade throughout the prosperous postwar era. Their efforts to mandate full employment responded to the continuing specter of technological unemployment as it hollowed out industries, communities, and skills. And as automation permeated factories and offices alike and worker anxieties rose, structural unemployment challenged the liberal-labor coalition to enact a major program for job skills training and one that assisted depressed communities impacted by unemployment.[1]

And yet despite this impressive record of accomplishment, the failure to enact a federal economic planning capability not only frustrated these indefatigable reformers but also exposed a fundamental reality borne out over their relentless campaign for greater economic justice. In a political economy in which technological progress was viewed as synonymous with economic prosperity, the very unpredictability of scientific discovery proved the most insurmountable obstacle to planning the future. Scientific discovery was simply an unpredictable force that, although enabled by ever-growing government research monies, resisted political constraints. As the physicist Niels Bohr famously remarked, "Prediction is extremely difficult. Especially about the future." If, for four decades, technology seemed always to be advancing too far and too fast for assembly line workers, telephone operators, and clerical employees, their political champions were similarly frustrated in their chase to anticipate the technological future. Moved by the unemployment he witnessed firsthand in the increasingly automated coal country of West Virginia in his 1960 campaign, Senator John Kennedy, who had seen his own region's textile industry hollowed out by more competitive and mechanized competition in the South, began to advocate for bolder solutions. Though as president he would sign his friend Paul Douglas's depressed areas legislation and Joe Clark's Manpower Development and Training Act, somehow these measures seemed insufficient as technology relentlessly transformed the economy.[2]

While the recommendations offered by the Automation Commission that Kennedy enabled may have finally met the threshold for boldness so often invoked by liberal-labor reformers, those recommendations came at a moment of economic and political transition. Economically, the unemployment rate fell to 3.8 percent in 1966. Less obviously, the report was issued just as US nonfarm productivity rates began their own steady descent. By the end of the next decade, economists more clearly saw how annual US productivity increases had significantly subsided as early as 1965. The trajectory of the postwar US industrial juggernaut was leveling off just as the commission delivered its findings. Politically, the growing demands of social movements protesting an escalating war, racial discrimination, and technological materialism challenged the postwar order. Amid swirling social unrest, the increasingly tenuous liberal-labor coalition found that a New Left movement was defining the terms of a new progressive agenda that often seemed far removed from the interests of the blue-collar worker. The very core of the liberal-labor coalition was having to rethink its own future.[3]

The commission had benefited from its own visionaries such as Bell and Reuther, and from forward-thinking business representatives such as Land and Watson. But although their many novel recommendations spoke to the future, it was a future removed from the immediate realities and thus the politics of the moment. The commission delivered its report as the liberal-labor coalition itself was splintering over the president's stubborn escalation of a war that was suddenly open to question and soon to ridicule. The growing public resonance of the antiwar platform stoked the momentum of the broader New Left agenda, with its own legacy stretching back to the Port Huron Statement and its own antiautomation bias. The emerging fractures within the liberal-labor coalition were apparent as early as the 1966 midterm election results. The stunning Democratic losses revealed that Johnson's impressive legislative track record came with its own heavy political price tag. Republicans won forty-seven seats in the House and three in the Senate. Symbolic of the demise of the liberal-labor coalition was the defeat of the inveterate New Dealer and friend of the working man Paul Douglas, who was convincingly trounced by former Bell and Howell Corporation CEO Chuck Percy. The trajectory of domestic reforms LBJ achieved in the thousand days since Kennedy's assassination was unceremoniously placed on pause. The New Left movement saw the midterm shellacking as further evidence of what it perceived as the stale policies of the old guard. Once the core of the liberal-labor alliance, the United States' blue-collar workforce now found itself sorting out its own priorities in the midst of spiraling domestic upheaval.[4]

The emerging tensions within the liberal-labor coalition were not lost on Walter Reuther, who sought to straddle the divide. His own ambivalence toward the multidimensional student movement and the insurgent new left wing of the Democratic Party was tempered by his desire to preserve labor's complicated relationship with the civil rights movement. His friend Martin Luther King Jr. provided him a pathway. King repeatedly joined the narratives of the civil rights, antiwar and antitechnology movements in his own inspirational sermons. In fact, King envisioned a reconstituted liberal-labor coalition springing from the feet of the automated future. He consistently preached that each of the often disparate protest movements of the moment was a product of the exploitive nature of modern technology. As he reminded his audience in his "Three Evils of Society" speech given at the Butler Street YMCA in Atlanta on May 27, 1967, the evils of war, racial discrimination, and unemployment were joined. As King observed, "A true revolution of values will soon look uneasily on the glaring contrast of poverty and wealth. With righteous indignation, it will look at thousands of working people displaced from their jobs with reduced incomes as a result of automation while the profits of the employers remain intact and say: 'This is not just.'" On April 4, 1968, however, King's healing voice was suddenly and tragically eliminated, and in the wake of his assassination the cycle of domestic protest and violence would only grow.[5]

Two convulsive years after King's death, and as social unrest spiraled violently, four students were killed by a National Guard unit on the very Kent State campus where A. H. Raskin had extolled the Automation Commission's findings. Then, one week after the Kent State massacre, Walter Reuther died in a fiery plane crash. The union movement, once the core of the liberal-labor coalition, found itself without what one eulogist described as "the crusader for a better world." In the void created by Reuther's sudden death, a seventy-six-year-old George Meany, who would become a welcome visitor at the Nixon White House, now sat atop an increasingly sclerotic union hierarchy. Nixon's appeal to a silent majority and the forgotten blue-collar worker illustrated the upside-down nature of a new politics. Indeed, as one observer of this tumultuous period noted, the nation seemed to be "coming apart," and the liberal-labor coalition's reforms suddenly seemed moored to a very distant past.[6]

Postindustrialism Arrives

After 1970, the reality of what Bell aptly termed the "post-industrial era" was becoming apparent in the shrinking number of manufacturing jobs as a percentage of overall US employment. Just as productivity growth began to contract in 1965, jobs in the manufacturing sector as a percentage of overall US jobs peaked that year at 28 percent. In his much-acclaimed *The Coming of Post-industrial Society*, Bell foretold the tectonic shifts in the US economy that were coming with the emergence of a new cadre of workers tied to computers. Bell noted that both industrial- and service-economy laborers would increasingly seek to take more control over their work environment and have more say over job content and quality. Bell had an obvious hand in the commission's own chapter on technology's impacts on the work environment, which concluded that work became more meaningful when employees could see their contribution to an overall process as well as a finished product. The commission acknowledged job design, the reorganization of the work process, and the recognition of human needs in the workplace as warranting more flexible work lives and offering flexible retirements. The commission's conclusions seemed prescient. In March 1972, the democratization of the workplace was at the heart of a contentious GM strike at its new Lordstown, Ohio, plant. The new, younger generation of union members asserted itself against the highly automated plant, not for the jobs eliminated, but because automation demeaned the dignity of existing jobs. The Lordstown strike was an uprising of alienated UAW members protesting the monotony of the modern assembly-line work conditions over which they had little say. The quality-of-work rationale they put forward as a negotiating position became known as "Lordstown Syndrome." It signaled a new attitude among rank-and-file union members concerned about their control over their work environment.[7]

Despite the attitudinal shift in union membership and the demands of the New Left agenda, the more traditional voices of the liberal-labor coalition would not go quietly into the night. A deep recession began in 1974, with unemployment nearing 8 percent. It would not recede below 6 percent for another decade. The economic malaise provided two veterans from the New Deal coalition with an opportunity to offer the notion of centralized federal economic planning once more. A stalwart FDR loyalist in his early days in state politics, California congressman Augustus Hawkins seized the moment by introducing the Equal Opportunity and Full Employment Act. Hawkins had already made his mark in the liberal-labor coalition as the principal author of Title VII of the Civil Rights Act, which prevented workplace discrimination based on race, color, national origin, sex, and religion. In his introductory remarks to his sweeping new legislative proposal, he noted that the earlier Employment Act of 1946 failed to acknowledge employment as a right. Full employment, Hawkins reminded his colleagues, should not be subordinate to those economic interests that sought to rationalize some tolerable level of unemployment. As he offered his own far-reaching economic planning measure, he noted that the outcome of the 1946 debate on the original Full Employment Act, almost thirty years before, was a far cry from what Senators Murray and Wagner had originally offered. Realizing that those political forces that gutted the earlier bill had only grown more emboldened, Hawkins sought to make full employment mandatory by legally protecting the right to have a job. The bill also established a Jobs Guarantee Office in a revamped federal Employment Service. The president would, in turn, be legally accountable for producing an annual national economic plan that met the goal of full employment, and failure to do so would automatically trigger programs to ensure employment to those who lost jobs in any future economic downturn.[8]

Noting that, at long last, it was time to fulfill the promise of jobs for all, another New Deal warrior, Hubert Humphrey, offered a similar bill in the Senate. Following his term as LBJ's vice president, and after his own failed presidential run in 1968, Minnesotans returned their beloved Humphrey to the Senate in 1970. In aligning with Hawkins's House measure, Humphrey's Senate version of the revitalized full employment legislation was joined by his old friend, the progressive Republican Jacob Javits. The Humphrey-Javits bill allowed US citizens to appeal to a US District Court if they felt they were being deprived of their right to a job. Moreover, it empowered the Federal Reserve to serve as a referee on the president's budgetary plans. The Hawkins and Humphrey bills allowed for what seemed another opportunity for the remnants of the liberal-labor coalition and its occasional New Left allies to achieve national economic planning and ensure full employment. Again, the battle over the federal government's role in planning for the economic future was joined, and these veterans of the liberal-labor coalition relished the fight.[9]

In the subsequent 94th (1975–76) and 95th (1977–78) Congresses, as unemployment remained unacceptably high, Humphrey and Hawkins joined together to offer identical bills that mandated 3 percent unemployment and the legally enforceable right to a job. The companion bills came to be both admiringly and disparagingly referred to simply as "Humphrey-Hawkins." When the Senate Banking Committee held hearings in May 1976, Chairman William Proxmire noted that although the bill was endorsed by every Democratic presidential candidate (except for George Wallace), it was less charitably described by critics as an engine of inflation, a budget buster, and an effort to repeal the business cycle. The fiscally conservative maverick Democrat from Wisconsin, known for exposing excessive defense contracts with his "golden fleece awards," was concerned that the bill could cost as much as $100 billion. Proxmire joined a growing list of senators signaling their intent to amend the legislation. Aware of what occurred in the debate during 1945–46, Humphrey, himself already weakened by cancer, was politically realist enough to know that in the political milieu of the late 1970s, compromise was necessary to enact this legislation.[10]

The bill's fate was further complicated by the election of fellow Democrat Jimmy Carter in 1976. The new president's lackluster support of a major labor law reform bill during the 95th Congress signaled just how much Democratic Party politics were in a state of transition, and how frayed the traditional liberal-labor alliance was. No longer could organized labor depend on a Democratic incumbent. Meanwhile, a reenergized business lobby that relied on its new grassroots power was resisting labor's attempt to roll back the right-to-work provisions of Taft-Hartley's so-called Section 14b. Though organized labor was simultaneously supporting the effort to enact Humphrey-Hawkins, it would spend more of its own waning political capital in an unsuccessful effort to invoke cloture on the more coveted repeal of 14b. The Humphrey-Hawkins bill became a lower union priority and had only tepid support from the Carter White House. While organized labor was thus distracted, the economic justice features of Humphrey-Hawkins carried special significance in the civil rights community, which threw its full weight behind the bill. As the March for Jobs and Freedom had amply demonstrated over a decade before, full employment was at the core of civil rights activism. But with Carter's reservations about many of the bill's provisions, and with business now riding the momentum of its victory in foiling the repeal of 14b, the Humphrey-Hawkins bill met a fate not all that dissimilar to that of its 1946 predecessor. The bill Carter finally signed on October 27, 1978, was viewed more as a symbolic tribute to its Senate author, who had died in January, than as a substantive commitment to full employment. Just as with the 1946 version, the federal economic planning provisions were again stripped from the bill. The crippled and compromised version of Humphrey-Hawkins passed on the very last day of the 95th Congress. But the neutered final version actually engaged the private sector in achieving its goals, which were not dissimilar, and in fact added little of substance, to those of the Employment Act of 1946.[11]

In an emotional signing ceremony in the East Room of the White House, Senator Muriel Humphrey (D-MN) and Congressman Hawkins flanked President Carter. An audience of two hundred, largely comprising labor representatives, civil rights activists, and urban reformers, gave the president and those who accompanied him a prolonged standing ovation. It was one last hurrah for the spirit of those who had sought for so long to protect the worker against the vagaries of technological advancement. In a few days, voters would again go to the polls in the midterm elections. Carter was feverishly campaigning for Democrats concerned about the continued white disaffection in union ranks, and the swelling proportion of unemployed blacks. He had just returned from Minnesota, where the former member of the Minnesota Democratic-Farmer-Labor Party and Democratic senator Walter Mondale, now Carter's vice president, was stumping for the embattled incumbent Wendell Anderson (D-MN). Lamenting the splintering of the party in the state, one party veteran said that Humphrey "was the grand uniter . . . he kept the whole thing together." Despite the president's and vice president's efforts, Minnesota, a state that was once a bastion of liberal-labor reformism, would elect two Republican senators.[12]

Back to the Future

Meanwhile, some three hundred miles away from the White House signing ceremony, in Lordstown, Ohio, automated robots were now a fixture on the shop floor. They had replaced the welders producing the Chevy Vega, a compact car that sought to compete with the new Japanese and German imports that continued to make steady gains in the US auto market in a suddenly more competitive global economy. Yet even the militant young Lordstown workers who rose up earlier in the decade to protest working conditions now seemed unfazed by the plant's increasing reliance on automation. Based on the agreements reached at the bargaining table with the UAW, any increases in GM productivity were now to be amply shared with the one million UAW members; this was the legacy of Reuther's hard-fought GAW. Moreover, the workers who lost their jobs were being retrained under the MDTA, some even to program their robotized replacements. A UAW worker who on average made $3.80 an hour in 1960 made $14 an hour in 1978, almost twice as much even when adjusted for inflation. The right to bargain had allowed workers to at least share some of the gains from increased productivity, even as the continued onslaught of new job-replacing machinery remained a stubborn reality and continued to take jobs, then livelihoods, then communities, and finally regions.[13]

During the 1984 presidential campaign, now Democratic presidential candidate Mondale, with his political moorings still firmly rooted in the Minnesota Democratic-Farmer-Labor Party, referred to the hollowing out of northeastern and upper Midwest manufacturing as creating the 1980s equivalent of the 1930s Dust Bowl. Mondale called the new phenomenon the Rust Bowl. New words now accompanied the emergence of a global economy, such as "outsourcing" and "supply chain," which, along with automation, became the insidious causes of the tragedy of what became popularly known as America's Rust Belt. By 1988, as globalization continued to take its toll in the steadily atrophying domestic manufacturing sector, Congress enacted the Worker Adjustment and Retraining Notification Act. It was the first worker protection law in two decades (since the enactment of the Occupational Safety and Health Act of 1970). The liberal-labor coalition, which once hoped for a national economic forecasting mechanism that could envision and thus mitigate the effects of future downturns in manufacturing employment, now settled for being politely warned, after a decision had already been made, that jobs were going to be eliminated—or, as Congressman Slack had put it two decades earlier, that they were "gone jobs."[14]

In the wake of the force of a globalized economy increasingly transformed by advancing technology and the birth of a new information economy, it seems doubtful that the creation of a national economic planning mechanism as envisioned by liberal-labor reformers in the period from 1921 to 1966, and later by Humphrey and Hawkins, could have significantly mitigated job loss. Indeed, it is hard to envision any forecasting capability that might have foreseen the rapid transformation of the US economy from its industrial dominance to a new economy characterized by the sheer speed with which information moves. The breathtaking scientific discoveries that would transform economic life later in the twentieth century were scarcely noticeable as the Automation Commission deliberated. In the ensuing decades after the commission's report, technology helped not only to fuel globalization but, in the course of doing so, to reshape the entire domestic economy. Not just industries and communities, but entire occupations and professions were eliminated with alarming speed. By 2018, Jeff Bezos, who pioneered a twenty-first-century technological transformation of the traditional retail space, noted that no business, however disruptive, was impervious to the later onslaught of a new wave of technological change. "I predict one day Amazon will fail. Amazon will go bankrupt," he warned. While meant to motivate employees to think anew, the statement was also a tacit acknowledgment that the challenge was similar to those faced by industrialists in the mid-twentieth century. Even the most visionary entrepreneurs realize that the next technological breakthrough could threaten their advantage if not eliminate the need for their business. The half-life of any enterprise, much less any skill, in the twenty-first century is now considered to be exponentially reduced from what it was just a few years before. Just as the unskilled laborer felt particularly vulnerable to technologies prior to World War II, and the skilled worker was increasingly victimized by computers in the post–World War II age of automation, in the next economy—and the next next economy—virtually no one could be assured a job of any permanence, much less a career.[15]

In an age about to welcome full-scale artificial intelligence and address the full dimensions of biotechnology, the political system will likely come under growing pressure to respond not only to anticipated new waves of unemployment, but also to continued changes in the very structure of our economic and social relationships. If the United States is to continue to respect both the value of technological progress and the dignity of work, then the sort of innovative policies that the liberal-labor coalition conceived as it chased after technological unemployment and then automation will be ever more in demand. Today, many of the very ideas that were shaped by these earlier reformers, from Robert Wagner to Philip Hart, remain as the tired and overused remnants of a progressive agenda from a half century ago. Even with alarming income disparity now well documented, and areas of the country falling behind amid technological innovation and global competitiveness, the technology-versus-jobs debate follows a pattern that would be familiar to those who engaged in that debate in the mid-twentieth century. Even as traditional conceptions of stable work and career-long professions have vanished in the new gig economy, discussion of the need for universal free education or even modest steps toward a basic income remain subject to the same well-worn critique that greeted the Automation Commission when it recommended such ideas in 1966. And, just as in the debate that ensued in the decades following the Harding Conference on Unemployment in 1921, today's generation of techno-optimists continues to insist that the next generation of discovery will alleviate any temporary jobs crisis.

When Republican senator Leverett Saltonstall rose in a mostly deserted Senate chamber on January 15, 1964, he asked the presiding officer if he might enter into the record a memoriam to the late President Kennedy, written by a student from Swampscott High School. Lola Kramarsky's tribute began by eloquently acknowledging "the gap between man's aspirations and human performance." Kennedy, she wrote, pulled us toward our ideals. What seemed to impress her the most was the way he recast the national agenda, setting an energetic tone and putting forward a bold set of goals. Perhaps, she averred, others could also take some comfort that Kennedy's successor, a celebrated creature of this very Senate chamber, was wasting little time in shaping the political landscape in order to enact many of the fallen president's forward-looking ideas, including his plan to form a presidential commission on automation.[16]

Indeed, that very commission proved to live up to its legislated charter. It developed a range of new ideas that sought to bridge the gap between technology and public policy beyond what seemed possible in the political moment. It did so with the same spirit that previously guided the determined corps of liberal-labor reformers as they sought to pull the United States toward the lofty goal of protecting those with jobs and those who lost them. Thus, the legacy of the liberal-labor coalition that chased after technology for over four decades was not that it failed, but rather that technology itself proved the continuing obstacle to any ultimate success. Though the liberal-labor coalition was unable to predict the technological future, its legacy was to blunt technology's worst effects at the height of US industrial prowess. And though their experience might well caution against any unreasonable expectations that politics can tame technology, their achievements derived from their unwavering pursuit of that goal.

NOTES

Introduction

1. Huthmacher, *Senator Robert Wagner*, 12.

2. Huthmacher, *Senator Robert Wagner*, 44–52, 137. The quote is taken from an interview with Anna M. Rosenberg, who served in various capacities in the Roosevelt administration, including as assistant secretary of defense, and was later was appointed to Lyndon Johnson's National Commission on Technology, Automation, and Economic Progress.

3. In the early nineteenth century, the Scottish botanist Robert Brown pondered the dilemma of the seemingly random motion of airborne pollen. Albert Einstein further refined Brown's theories. Wiener used mathematical methods to predict and explain them in his paper entitled "Generalized Harmonic Analysis" in the Swedish journal *Acta mathematica*. Wiener, *Cybernetics*, 1–29; Conway and Siegelman, *Dark Hero*, 48–55, 63–67, 138–43.

4. The term "liberal-labor coalition" is used here to describe the reform-oriented coalition that emerged full blown during the New Deal and that is characterized by Jefferson Cowie in *The Great Exception*, 9. It is also used in Zieger, *For Jobs and Freedom*, 186, 190, 203. In his still much-cited history of the Employment Act of 1946, Stephen Bailey abbreviates it as the "lib-lab" coalition; see Bailey, *Congress Makes a Law*.

5. Bix, *Inventing Ourselves*, 7; Hounshell, *American System*; Pursell, *Technology in Postwar America*; Zieger, *For Jobs and Freedom*; Lichtenstein, *Labor's War at Home*; Woirol, *Technological Unemployment*; Cowie, *Great Exception*; Zelizer, *Fierce Urgency of Now*; Perlstein, *Before the Storm*; Shermer, *Sunbelt Capitalism*.

6. Jacobs, "Uncertain Future," 155. For additional insight into the major players in the liberal-labor coalition, see Zelizer, *Fierce Urgency of Now*; Shales, *Great Society*; Milkis and Mileur, *Great Society*.

7. Sautter, *Three Cheers*, 42–62, 73–80; Stricker, *American Unemployment*, 32–38; Salvatore, *Eugene V. Debs*, 189–243; Prout, *Coxey's Crusade for Jobs*, 86–109; Postel, *Populist Vision*, 228–39; O'Donnell, *Henry George*, 47–50, 209–10.

8. Grant, *Forgotten Depression*, 67–80.

9. Lorant, "Technological Change"; William Green, "Labor versus Machines," *New York Times*, June 30, 1930; Smiley, *Rethinking the Depression*, 4–5; Pursell, *Technology in Postwar America*, 1–2; Hounshell, *American System*, 217–61; Sumner H. Slichter, "The Price of Industrial Progress," *New Republic*, February 8, 1928, 316–17; Field, *Great Leap Forward*, 19; Tugwell, *Industry's Coming of Age*; Slichter, "Current Labor Policies."

10. President Franklin D. Roosevelt, "1944 State of the Union Address," Franklin D. Roosevelt Library and Museum, January 11, 1944, http://www.fdrlibrary.marist.edu/archives/stateoftheunion.html.

11. Zachary, *Endless Frontier*, 260.

12. Phillips-Fein, "Conservatism"; Brinkley, *End of Reform*, 256–64; Critchlow, *Conservative Ascendancy*, 6–23. McGirr, *Suburban Warriors*, 37.

13. Katznelson, Greiger, and Kryder, "Limiting Liberalism." Katznelson later expanded this thesis in *Fear Itself*, suggesting that the triumph of progressive liberalism was politically enabled temporarily by a southern Democratic faction that soon departed when the reformist impulse of the liberal-labor coalition turned to civil rights. See also Lichtenstein, "Class Politics"; Shermer, "Counter-organizing the Sunbelt."

14. President Lyndon B. Johnson, "Remarks upon Signing Bill Creating the National Commission on Technology, Automation, and Economic Progress," American Presidency Project, August 19, 1964, https://www.presidency.ucsb.edu/documents/remarks-upon-signing-bill-creating-the-national-commission-technology-automation-and. Manufacturing jobs in the United States in January 1920 were estimated to be 9.2 million, and by January 1969 stood at 18.7 million. Manufacturing jobs data from 1920 to 1970, and from 1939 to the present, are available through Federal Reserve Economic Data, Federal Reserve Bank of St. Louis; see both "Production Worker Employment, Manufacturing, Total for United States," accessed October 28, 2021, https://fred.stlouisfed.org/series/M081FBUSM175NNBR; and "All Employees, Manufacturing," accessed October 28, 2021, https://fred.stlouisfed.org/series/CEU3000000001.

15. Senate Temporary National Economic Committee, *Investigation of Concentration of Economic Power*, April 26, 1940, 17244.

16. As Carl Benedikt Frey points out, the distinctions between the impacts of enabling versus replacing technologies on wages and employment remained largely unnoticed, even through the groundbreaking work of Jan Tinbergen in the 1970s; see Frey, *Technology Trap*, 13–16. Also see Acemoglu and Restrepo, "Automation and New Tasks," which concludes, "Our evidence and conceptual approach support neither the claims that the end of human work is imminent nor the presumption that technological change will always and everywhere be favorable to labor" (27). See also the testimony of Daron Acemoglu in House Committee on the Budget, *Hearing on Machines*, 38–48.

17. Lichtenstein, *Walter Reuther*, 290.

18. Diebold, *Automation*, 158–65; Joint Committee on the Economic Report, *Hearings on Automation*, 6–48.

19. Frey, *Technology Trap*, 1–5.

20. Cowie, *Great Exception*, 167–70; Cowie and Salvatore, "Long Exception"; Weir, *Politics and Jobs*, 62–98, 163–80; Kuhn, *Hard Hat Riot*.

21. National Commission on Technology, Automation, and Economic Progress, *Technology and the American Economy*, xii; Bell, "Government by Commission," 9.

22. Bureau of Labor Statistics Top Picks, "Unemployment Data 1929–2020," accessed February 16, 2021, https://data.bls.gov/cgi-bin/surveymost?bls. As early as 1971, Robert Solow joined Edward Denison and George Perry in observing that although in the period from 1948 to 1970 manufacturing employment continued to go up, productivity rates were declining; see Perry, Denison, and Solow, "Labor Force Structure." In 1979, the Brookings Institute began a series of analyses that revealed that labor productivity, which grew at an annual rate of 2.57 percent a year from

1948 to 1968 in the nonfarm sector of the economy, fell to 1.64 percent in the period 1968–73; see Bailey, "Productivity Growth Showdown." Robert Gordon notes the amazing growth in productivity between 1920 and 1970 and its slowing thereafter; see Gordon, *Rise and Fall*, i–ii; Field, *Great Leap Forward*, 19.

23. Noble, *Forces of Production*, xiii, 43.

1. Voices for the Unemployed

1. Davis, *Iron Puddler*, 56–58, 90–107; Misa, *Nation of Steel*, 4–5, 55–57, 256–57; Landes, *Unbound Prometheus*, 89–95, 250–52.

2. James L. Davis, "Unemployment in the United States," *Congressional Record*, 67th Cong., 1st sess., August 16, 1921, vol. 61, pt. 5, 5036–37; Grant, *Forgotten Depression*, 5–6.

3. "Senator McCormick Served Citizens in Many Capacities," *Chicago Daily Tribune*, February 26, 1925; Stone, "Two Illinois Senators"; Senator Joseph McCormick, "Unemployment in the United States (S.J. Res. 126)," *Congressional Record*, 67th Cong., 1st sess., August 5, 1921, vol. 61, pt. 5, 4695; Davis, "Unemployment"; "U.S. Has More Than 5,700,000 Jobless," *Chicago Daily Tribune*, August 17, 1921.

4. "Congressman Meyers London," *Congressional Record*, 67th Cong., 1st sess., May 11, 1921, vol. 61, pt. 4, 1333; Clements, *Life of Herbert Hoover*, 131–33; Jeansonne, *Herbert Hoover*, 156–57; "The Facts as to Employment," *New York Times*, September 15, 1921; "President Opens Conference to Aid the Unemployed," *New York Times*, September 27, 1921; Gaddis, "Herbert Hoover," 24–25; "Harding Formally Opens Conference on Unemployment," *Atlanta Constitution*, September 27, 1921; *Report of the President's Conference*, 29.

5. "President Warren Harding Address at the Opening of the Conference on Unemployment," September 21, 1921, American Presidency Project, https://www.presidency.ucsb.edu/documents/address-the-opening-the-conference-unemployment-called-secretary-commerce-hoover.

6. "President Warren Harding Address"; Sautter, *Three Cheers*, 130–31; "Get Busy on Unemployment," *Los Angeles Times*, September 28, 1921.

7. Grant, *Forgotten Depression*, 199–202; Remini, *House*, 299. The Coolidge quote appears in Kennedy, *American People*, 33.

8. Reich, "From the Spirit of St. Louis," 353–56; Allen, *Only Yesterday*, 141; "There's Magic in the Air," *Collier's*, April 7, 1928. Also see Bix, *Inventing Ourselves*, 11, 314n6.

9. Bix, *Inventing Ourselves*, 9–12; Hounshell, *American System*, 305; House Judiciary Committee, *Unemployment*, 47–48; William Green, "Labor versus Machines," *New York Times*, June 1, 1930.

10. Hounshell, *American System*, 306–10.

11. Green, "Labor versus Machines." AFL president William Green used Federal Reserve and Employment Service Bureau numbers to compile these estimates of manufacturing and productivity growth. Subsequent analysis showed these numbers to be an accurate characterization of the 1920s growth economy; see Gene Smiley, "The U.S. Economy in the 1920s," Encyclopedia of the U.S. Economy, accessed August 7, 2021, https://eh.net/encyclopedia/the-u-s-economy-in-the-1920s/. More recently, Alexander J. Field has noted that "manufacturing contributed almost

all . . . of the total growth factor productivity in the U.S. private nonfarm economy between 1919 and 1929"; see Field, "Technological Change," 213 (specifically the annual productivity growth rate was 5.12 percent annually); Field calls the 1920s a transitional decade in US economic history and "unique in the history of U.S. manufacturing"; see Field, *Great Leap Forward*, 46–47; Lorant, "Technological Change," 243–44.

12. "Unions Now to Demand 'Higher Social Wages,'" *Washington Post*, July 12, 1927; Green, "Labor versus Machines"; Robert H. Zieger, *Republicans and Labor*, 250; "William Green Is Dead at 84; Headed AFL since 1924," *New York Times*, November 22, 1952.

13. Stewart, "Wastage of Men"; Bernstein, *Lean Years*, 50.

14. Sumner Slichter, "The Price of Industrial Progress," *New Republic*, February 8, 1928, 316–17; Roediger and Foner, *Our Own Time*, 237–39; "Five Day Week Becomes a Vivid Issue," *New York Times*, October 17, 1926.

15. As president of the AFL, Green proposed the suspension of immigration for a period of ten years as part of an overall plan to alleviate unemployment. "Labor Plans Drive on Unemployment," *Boston Globe*, February 21, 1929; William Green, "Prosperity and Unemployment," *Atlanta Constitution*, September 1, 1929; Green, "Labor versus Machines"; Sautter, *Three Cheers*, 220–22; Bix, *Inventing Ourselves*, 14–15; Field, "Technological Change," 213; Slichter, "Price of Industrial Progress," 317.

16. Palmer, *Twenties in America*, 177. Several sources address how the political system failed to recognize unemployment throughout the 1920s; these include Patterson, *New Deal and States*, 3–6; Sautter, *Three Cheers*, 2–3; Keyssar, "Unemployment"; Sautter, "Government and Unemployment"; "Statistics Relative to Unemployment," *Congressional Record*, 70th Cong., 1st sess., March 5, 1928, vol. 69, pt. 4, 4067–69; Huthmacher, *Senator Robert Wagner*, 5–15, 42–43; Von Drehle, *Triangle*, 200–202; Stein, *Triangle Fire*, 20.

17. Huthmacher, *Senator Robert Wagner*, 57–69; "Hoover Lays Plans for Labor Support," *Washington Post*, September 16, 1928; "Statistics Relative to Unemployment," 4067–69; "Statistics Relative to Unemployment," 70th Cong., 1st sess., February 23, 1928, vol. 69, pt. 3, 3419; Sautter, *Three Cheers*, 245.

18. "Wagner Declares Unemployment Rife; Senate Asks Inquiry," *New York Times*, March 6, 1928; "Presidential Campaigns," *Boston Globe*, March 6, 1928; Finan, *Alfred E. Smith*, 202; Huthmacher, *Senator Robert Wagner*, 12–17, 52–53; 107–8.

19. "Unemployment Conditions," *Congressional Record*, 70th Cong., 1st sess., April 20, 1928, vol. 69, pt. 6, 6839–43; Huthmacher, *Senator Robert Wagner*, 58–62; Bernstein, *Lean Years*, 267.

20. "Senator Robert Wagner, S. 4157," *Congressional Record*, 70th Cong., 1st sess., April 20, 1928, vol. 69, pt. 6, 6811; "Unemployment Conditions"; "Progress in American Labor Legislation," 50; Bernstein, *Lean Years*, 267; Bix, "Inventing Ourselves," 158–59.

21. "Senator Robert Wagner, S. 4158," *Congressional Record*, 70th Cong., 1st sess., April 20, 1928, vol. 69, pt. 6, 6811; "Unemployment Conditions"; Huthmacher, *Senator Robert Wagner*, 59–63; Sautter, *Three Cheers*, 247.

22. "Unemployment Conditions"; "Senator Robert Wagner, S. 4307," *Congressional Record*, 70th Cong., 1st sess., May 1, 1928, vol. 69, pt. 7, 7502.

23. "Unemployment Conditions"; Huthmacher, *Senator Robert Wagner*, 59–63; Brewer, "Evolution and Alteration," 121–23.

24. Slichter, "Price of Industrial Progress," 316–17; Woirol, *Technological Unemployment*, 223–28; Gordon, *Rise and Fall*, 269–71; "Wagner Asks Speed on Jobless Bills," *New York Times*, May 25, 1930; Huthmacher, *Senator Robert Wagner*, 68–74.

25. Barnard, *Independent Man*, loc. 109–34 of 8683, Kindle; Amsterdam, "Civic Welfare State," 43–52; "Senator Couzens of Michigan Dies," *New York Times*, October 23, 1936; US Senate, *Report Pursuant to S. Res. 219*, 3–11.

26. Kates, "Editor, Publisher, Citizen Soldier"; "Compulsory Unemployment Insurance," *Congressional Record*, 70th Cong., 1st sess., March 26, 1928, vol. 69, pt. 5, 5380.

27. US Senate, *Report Pursuant to S. Res. 219*, 3–11; Senate Committee on Education and Labor, *Causes of Unemployment*, 212–44. "The Wisconsin idea" was a term used to describe the integral relationship between the University of Wisconsin and progressive reformers in the Wisconsin legislature, including Robert La Follette; see Hoeveler, *John Bascom*, loc. 83 of 5542, Kindle.

28. "Unemployment in the United States," *Congressional Record*, 71st Cong., 2nd sess., March 3, 1930, vol. 72, pt. 5, 4594–616; Huthmacher, *Senator Robert Wagner*, 78–80; Sautter, *Three Cheers*, 249.

29. "Unemployment in the United States."

30. Palmer, *Twenties in America*, 177; Stricker, *American Unemployment*, 43–49; Huthmacher, *Senator Robert Wagner*, 70–73; House Judiciary Committee, *Unemployment*, 22–25; biography of George S. Graham, History, Art & Archives, US House of Representatives, accessed August 7, 2021, https://history.house.gov/People/Listing/G/GRAHAM,-George-Scott-(G000354)/.

31. Green, "Prosperity and Unemployment"; House Judiciary Committee, *Unemployment*, 26–40, 43–50; biography of Emanuel Celler, History, Art & Archives, US House of Representatives, accessed August 7, 2021, https://history.house.gov/People/Listing/C/CELLER,-Emanuel-(C000264)/; biography of Hatton William Sumners, History, Art & Archives, US House of Representatives, accessed August 7, 2021, https://history.house.gov/People/Listing/S/SUMNERS,-Hatton-William-(S001072)/.

32. Bernstein, *Lean Years*, 247–52; Whyte, *Hoover*, 345; Jeansonne, *Herbert Hoover*, 84–87, 162–70; E. Roy Weintraub, "Neoclassical Economics," Concise Encyclopedia of Economics, accessed August 7, 2021, https://www.econlib.org/library/Enc1/NeoclassicalEconomics.html.

33. "Text of Hoover Speech" (reprint of speech to the American Federation of Labor, Boston, October 6, 1930), *New York Times*, October 7, 1930; Bix, *Inventing Ourselves*, 44; "U.S. Prosperity Is Mere Nibble," *Christian Science Monitor*, May 15, 1929; Weber, *Homing Pigeons*, 117–18; Vought, *Bully Pulpit*, 216–17; "Jobs vs. Machinery," *Washington Post*, March 28, 1930; "Groundwork Is Laid for Hoover Parley," *New York Times*, August 25, 1932.

34. "Hoover Hails Labor's Aid in Unemployment Crisis," *New York Times*, October 7, 1930; "Text of Hoover Speech"; Green, "Labor versus Machines"; Senate Committee on Education and Labor, *Causes of Unemployment*, 186.

35. "Ewan Clague, a Labor Official," *New York Times*, April 15, 1987; "Report of the Advisory Committee," 16–22.

36. "Report of the Advisory Committee," 23; Paul H. Douglas, "Technological Unemployment," *American Federationist*, August 1930, 932–50 (quotation on 936).

37. S. 5776 passed the Senate on April 2, 1930, and the House on May 19, 1930, and was signed into law on July 7, 1930. Sautter, *Three Cheers*, 248; "Unemployment Study," *Wall Street Journal*, December 25, 1931; "Hoover Receives Unemployment Bill," *Christian Science Monitor*, July 3, 1930; "Labor Statistics by the End of the Year," *New York Times*, July 3, 1930.

38. Jeansonne, *Herbert Hoover*, 222–28; Huthmacher, *Senator Robert Wagner*, 77–78.

39. S. 3060 was introduced on January 6, 1930; voted favorably in the Senate on April 8, 1930; voted favorably in the House on June 26, 1930; and pocket vetoed by Hoover on March 3, 1931. "House of Representatives Floor Debate on S. 3060," *Congressional Record*, 71st Cong., 3rd sess., February 23, 1931, vol. 74, pt. 6, 5774; "Hoover Kills Wagner Bill by Pocket Veto," *Baltimore Sun*, March 8, 1931; Herbert Hoover, "Statement on the Appointment of John R. Alpine to the United States Employment Service," March 13, 1931, American Presidency Project, https://www.presidency.ucsb.edu/documents/statement-the-appointment-john-r-alpine-the-united-states-employment-service; Huthmacher, *Senator Robert Wagner*, 78–80.

40. "Finds Labor Unhurt by Technical Gains: Report to Chemical Engineers Says It Is Only a Minor Cause of Unemployment," *New York Times*, December 11, 1931; "Biographical Note," Frederick C. Croxton Papers, Herbert Hoover Presidential Library and Museum, accessed May 16, 2019, https://hoover.archives.gov/research/manuscript-collections/croxton.

41. Leuchtenberg, *Herbert Hoover*, 91, 136–38; Whyte, *Hoover*, 499–503; Dickson and Allen, *Bonus Army*, 95–99; William Green, "Labor's Plan for Recovery: A Five-Day Week for Workers," *New York Times*, July 17, 1932; "Beer Works Plan Advised by Green," *New York Times*, June 27, 1932.

42. Thompson and Whelpton, "Population of the Nation," 33; Gay and Wolman, "Trends in Economic Organization," 234–37; Ogburn, "Influence of Invention," 128–29; Ogburn, *Social Change*, 233–65; "In Memoriam: Leo Wolman"; Hurlin and Givens, "Shifting Occupational Patterns," 310; "Committee Findings," xv.

43. "Remarks of James Davis," *Congressional Record*, 76th Cong., 1st sess., January 31, 1933, vol. 76, pt. 3, 2985–92.

44. "Remarks of James Davis," 2994.

2. Taming Technology

1. "Here's What Ails the U.S.A.," and "Capital Eyeing Roosevelt for Word of Plans," *Chicago Daily Tribune*, January 2, 1933; Irving Bernstein, "Americans in Depression and War," US Department of Labor, accessed August 8, 2021, https://www.dol.gov/general/aboutdol/history/chapter5.

2. Badger, *First One Hundred Days*, ix–xiii, 91–93; Godin, "Innovation without the Word," 286; McElvaine, *Great Depression*, 174–75, 287–90; Cowie, *Great Exception*, 100–101; Stershner, "Victims of the Great Depression."

3. "National Employment System," *Congressional Record*, 73rd Cong., 2nd sess., June 1, 1933, vol. 77, pt. 5, 4766–83; Alter, *Defining Moment*, 291–93; Kennedy, *American People*, 144–47; An Act to Provide for a National Employment System, PL 73-30, 73rd Cong., 1st sess., June 6, 1933.

4. Federal Employment Relief Act, PL 73-15, 73rd Cong., 1st sess., May 12, 1933; Franklin D.

Roosevelt, "Statement on Signing the Unemployment Relief Bill," May 12, 1933, American Presidency Project, https://www.presidency.ucsb.edu/documents/statement-signing-the-unemployment-relief-bill; Rauchway, *Winter War*, 138–39; Badger, *First One Hundred Days*, 58–63; Kennedy, *American People*, 168–77.

5. Kennedy, *American People*, 260; Dray, *Power in a Union*, 128; Bix, *Inventing Ourselves*, 108–10.

6. "Senator Black Remarks on Six Hour Day and Thirty Hour Week," *Congressional Record*, 72nd Cong., 2nd sess., February 17, 1933, vol. 76, pt. 4, 4304–15.

7. Milton, *Politics of U.S. Labor*, 38–73; Roediger and Foner, *Our Own Time*, 243–56; "To prevent interstate commerce . . .," S. 5267, 72nd Cong., 2nd sess., December 8, 1932; Hunnicutt, "Kellogg's Six-Hour Day"; Walker, "Share the Work Movement," 16–17.

8. "Senator Black Remarks"; "Five Day Week and Six Hour Day," *Congressional Record*, 73rd Cong., 1st sess., April 5, 1933, vol. 77, pt. 2, 1291–97; "Senate Vote on Thirty Hour Week Bill Today," *Chicago Daily Tribune*, April 6, 1933; Kennedy, *Freedom from Fear*, 150–51: Dewey L. Fleming, "Minimum Wage Is Planned for 30 Hour Work Week," *Baltimore Sun*, May 7, 1933; W. R. Francis, "Short Work Week Plan Wins," *Los Angeles Times*, April 7, 1933; Huthmacher, *Senator Robert Wagner*, 6; "Two Roosevelt Aides Back 30-Hour Bill," *Baltimore Sun*, April 12, 1933.

9. "Miss Perkins Asks Industry Control," *New York Times*, April 19, 1933; "Swope Joins Green for Thirty Hour Week," *New York Times*, April 27, 1933; Rosen, *Roosevelt*, loc. 1919–2034 of 7096, Kindle.

10. Cole and Ohanian, "New Deal Politics," 785.

11. NIRA was signed into law on June 16, 1933, and was ruled unconstitutional by the Supreme Court on May 27, 1935. Ohl, *Hugh Johnson*, 102, 266; Dray, *Power in a Union*, 416–21; Milton, *Politics of U.S. Labor*, 25–37; Kennedy, *Freedom from Fear*, 298–302.

12. "Back to Work," *Washington Post*, June 23, 1933; William MacDonald, "Miss Perkins Looks Ahead," *New York Times*, May 27, 1934.

13. Tugwell, "Theory of Occupational Obsolescence," 181, 199, 212–25.

14. "Miss Perkins Looks Ahead."

15. Stuart Chase, "Machines Winning in the Battle for Jobs," *Washington Post*, August 5, 1934; Stuart Chase and Charles W. Hurd, "Works Relief Disputes Laid before Roosevelt; He Reviews the Program," *New York Times*, September 12, 1935.

16. Current analysis shows that estimates at the time were essentially providing an accurate picture of the disparity between GNP growth rates and unemployment. See Federal Reserve Economic Data, "Gross National Product," Federal Reserve Bank of St. Louis, accessed May 11, 2020, https://fred.stlouisfed.org/series/A001RP1A027NBEA; Federal Reserve Economic Data, "Unemployment Rate for the United States," Federal Reserve Bank of St. Louis, accessed May 12, 2020, https://fred.stlouisfed.org/series/M0892AUSM156SNBR; "Counting the Jobless," *Baltimore Sun*, May 2, 1935. The *New York Times* cited International Labor Organization numbers that reported that in April 1934, the number of US jobless was approximately 10,900,000, and in April 1935 it stood at 11,500,000; see "Year Fails to Cut Jobless in World," *New York Times*, July 11, 1935; "9,710,000 Jobless, an

Increase Found," *New York Times*, June 26, 1935. In the years between the census data from 1930 and 1940, unemployment numbers were calculated not by counting the number of unemployed, but by deducting the total of those employed from the total actually available to work. Amid the number of independent surveys, it turns out that the BLS numbers roughly approximate those of the AFL; see table 2 in Bureau of Labor Statistics, "Technical Note," 1948, https://www.bls.gov/opub/mlr/1948/article/pdf/labor-force-employment-and-unemployment-1929-39-esti mating-methods.pdf.

17. "Five States Have Enacted Bills for Unemployment Insurance," *Washington Post*, July 8, 1935; House Subcommittee of the Committee on Ways and Means, *Unemployment Insurance*, 28.

18. House Ways and Means Committee, *Economic Security Act*, 5–27. For a discussion of dramatically rising productivity rates during the Great Depression, see Field, *Great Leap Forward*, 35–40.

19. "Introduction S.1130," *Congressional Record*, 74th Congress, 1st sess., January 17, 1935, vol. 79, pt. 1, 549–55; "Old Age Pension and Insurance Plan Legislation Introduced in Congress and Endorsed by Roosevelt for Immediate Passage," *Christian Science Monitor*, January 17, 1935; "Article by Senator Wagner," *Congressional Record*, 74th Cong., 2nd sess., February 20, 1936, vol. 80, pt. 3, 2415; "Social Security in America: The Role of the Federal Government in Unemployment Compensation," Social Security Administration, accessed August 8, 2021, https://www.ssa.gov/his tory/reports/ces/cesbookc5.html; Price, "Unemployment Insurance"; House Ways and Means Committee, *Economic Security Act*, 28; "Five States Have Enacted Bills"; Social Security Administration, "Unemployment Insurance," accessed August 8, 2021, https://www.ssa.gov/policy/docs/progdesc/sspus/unemploy.pdf; Haber and Joseph, "Unemployment Compensation"; Kates, "Editor, Publisher, Citizen Soldier"; Sautter, *Three Cheers*, 334–38. For a description of the difference between the new insurance system under the SSA and the expired FRE relief system, see Rosen, *Roosevelt*, loc. 3045–23 of 7096, Kindle.

20. A. L. A. Schechter Poultry Corporation v. United States, Oyez, accessed August 8, 2021, https://www.oyez.org/cases/1900-1940/295us495; "Senator Wagner Explanatory Remarks S. 1958," *Congressional Record*, 74th Cong., 1st sess., February 21, 1935, vol. 79, pt. 3, 2371–72; "Labor's Pet Bills Run into Trouble," *New York Times*, May 2, 1935.

21. "Worker's Unemployment, Old Age, and Social Insurance Bill History and Status of S. 2827," *Congressional Record*, 74th Cong., 1st sess., April 3, 1935, vol. 79, pt. 5, 4971; "NRA Codes, Address by Senator Wagner," *Congressional Record*, 74th Cong., 1st sess., March 5, 1934, vol. 78, pt. 4, 3678–79; "Lundeen Investigation Denied," *New York Times*, September 17, 1940; Albert Eisele, "Death of Senator from Minnesota Still Shrouded in Mystery," *Minnesota Post*, September 3, 2009, https://www.minnpost.com/politics-policy/2009/09/death-senator-minnesota-still-shrouded-mystery.

22. "Machines Winning in the Battle for Jobs"; Bernard Kilgore, "Permanent Unemployment," *Wall Street Journal*, September 13, 1935; "30-Hour Work Week Endorsed by Roosevelt," *Baltimore Sun*, April 13, 1933; "The World's Business," *Christian Science Monitor*, November 7, 1935.

23. "Unemployment amid Recovery: Vast Riddle," *New York Times*, April 26, 1936; "Continued Unemployment," *Daily Boston Globe*, April 17, 1936; "Two Year Job Rise

Is Put at 5,413,000," *New York Times*, April 14, 1936; "Unemployment in U.S. Reaches Recovery Low," *Washington Post*, October 14, 1936; "Text of President's Talk before New York Party Assembly," *Los Angeles Times*, April 26, 1936. The Federal Reserve compiled numbers tracking both the BLS and other unemployment surveys. Federal Reserve Economic Data, "Unemployment Rate for the United States"; House Committee on Labor, *Hearings on H. Res. 49*, February 17, 1936, 44; "March Unemployment Put at 12,184,000," *Wall Street Journal*, April 30, 1936; "National Chamber Plans Wide Survey to Produce Jobs," *New York Times*, April 28, 1936.

24. "John Lesinski," Political Graveyard, accessed June 23, 2020, http://political graveyard.com/bio/leonardo-lessler.html#R9M0J4259; "Obituary: Vincent Palmisano," *Baltimore Sun*, March 6, 1953; "Palmisano's Body Found at Baltimore," *Washington Post*, March 5, 1953; House Committee on Labor, *Hearings on H. Res. 49*, February 17, 1936, 43–50.

25. Woirol, "Machine Taxers"; see also Woirol, *Technological Unemployment*, 36; *John Lesinski Memorial Services*; "John Lesinski"; "Technotax Urged for Machines," *Los Angeles Times*, April 14, 1935; House Committee on Labor, *Hearings on H. Res. 49*, February 17, March 2, 1936, 43–50, 79.

26. Bix, *Inventing Ourselves*, 171; "Requesting the Secretary of Labor to Compile a List of the Labor Saving Devices; and for Other Purposes," House Committee on Labor, *Report to Accompany H. Res. 49*, 2–3; "House Resolution 49," *Congressional Record*, 74th Cong., 2nd sess., June 20, 1936, vol. 80, pt. 10, 10579, 10659; House Committee on Labor, *Hearings on H. Res. 49*, February 20, 1936, 51–65. In discussing a techno-tax, Harry Jerome of the National Bureau of Economic Research speculated that "it is probable that sometimes the desire to reduce points of friction with workers leads employers to substitute docile machines for more vociferous human labor." See Jerome, *Mechanization in Industry*, 353, National Bureau of Economic Research, https://www.nber.org/books-and-chapters/mechanization-industry.

27. Corrington Gill, "The Case for Relief in an Era of Recovery," *New York Times*, August 29, 1937; "Ghost over the White House," *New York Times*, November 1, 1935; "He Reviews the Program," *New York Times*, September 12, 1935; Bix, *Inventing Ourselves*, 57–71; Stanley Lebergott, "Labor Force, Employment, and Unemployment, 1929–39: Estimating Methods," US Bureau of Labor Statistics, July 1948, https://www.bls.gov/opub/mlr/1948/article/labor-force-employment-and-unemployment-1929-39-estimating-methods.htm; Corrington Gill, "Job Survey Launched by Government," *Los Angeles Times*, June 14, 1936; Corrington Gill, "WPA Studies 'Foe' the Swift Machine," *New York Times*, May 31, 1936.

28. Weintraub, "Unemployment and Increasing Productivity"; "New Machines Destroy Jobs, WPA Reports," *Christian Science Monitor*, March 29, 1937. For a list of the sixty reports completed by the WPA National Research Project, see Gill, *Unemployment and Technological Change*; also see Weintraub, *Work and Publications*.

29. Newman, *Hugo Black*, 215–19; Samuel, "Troubled Passage"; Leuchtenberg, *Franklin D. Roosevelt*, 177–191; Patricia Waiwood, "The 1937–38 Recession," Federal Reserve History, accessed August 7, 2020, https://www.federalreservehistory.org/essays/recession-of-1937-38.

30. Samuel, "Troubled Passage"; Fleck, "Democratic Opposition"; Alter, *Defining Moment*, 293; Kennedy, *Freedom from Fear*, 339–48; Newman, *Hugo Black*, 216.

31. "Miss Perkins Asks Industry Control"; "President Favored Labor Says AFL," *New York Times*, October 18, 1936; "Wage and Hour Bill," *Congressional Record*, 75th Cong., 1st sess., July 29, 1937, vol. 81, pt. 7, 7847–48; "Pass the Wage and Hours Bill," *Congressional Record*, 75th Cong., 1st sess., December 16, 1937, vol. 82, pt. 2, 1672–73; Howard D. Samuel, "Troubled Passage: The Labor Movement and the Fair Labor Standards Act," *Monthly Labor Review*, December 2009, 32–37; President Franklin D. Roosevelt, fireside chat, June 24, 1938, American Presidency Project, https://www.presidency.ucsb.edu/documents/fireside-chat-14.

32. Dallek, *Franklin D. Roosevelt*, 486; Roosevelt, fireside chat, June 24, 1938.

33. Dallek, *Franklin D. Roosevelt*, 288–91; Gill, *Unemployment and Technological Change*, 16–17; Franklin D. Roosevelt, "Annual Message to Congress," January 3, 1940, American Presidency Project, http://www.presidency.ucsb.edu/ws/index.php?pid=15856.

34. President Franklin D. Roosevelt, "Message to the Congress on the Concentration of Economic Power," April 29, 1938, Pepperdine School of Public Policy, https://publicpolicy.pepperdine.edu/academics/research/faculty-research/new-deal/roosevelt-speeches/fr042938.htm; US Senate, *Report to Accompany S.J. Res. 300*; Ray Hill, "The Man from Wyoming," *Knoxville Focus*, April 17, 2016, https://knox-focus.com/archives/this-weeks-focus/man-wyoming-senator-joseph-c-omahoney/; "Problems of Employment," *Congressional Record*, 74th Cong., 2nd sess., March 26, 1936, vol. 80, pt. 4, 4375–76.

35. John L. Hess, "Isador Lubin, 82, Dies," *New York Times*, July 8, 1978; Huthmacher, *Senator Robert Wagner*, 68, 83, 110.

36. Senate Temporary National Economic Committee, *Investigation of Concentration*, December 1, 1938, i–67.

37. Senate Temporary National Economic Committee, *Investigation of Concentration*, April 6, 19, 25, 1940, 16209–67, 16760–96, 17122–40; Bix, *Inventing Ourselves*, 227–32.

38. Senate Temporary National Economic Committee, *Investigation of Concentration*, December 2, 1941, 123–65.

39. Testimony of Isador Lubin, in Senate Temporary National Economic Committee, *Investigation of Concentration*, April 26, 1940, 17242–65; statement of Lewis Lorwin and John Blair, in *Technology in Our Economy*, xi–xiii, 219–20.

3. Technology's Triumph

1. See data on unemployment and gross domestic product (GDP) compiled by Federal Reserve Economic Data, Federal Reserve Bank of St. Louis, accessed September 7, 2020, https://fred.stlouisfed.org/series/A191RP1A027NBEA. Up until 1991, gross national product (GNP) was the standard measurement of economic output; GNP measures total output of a nation's residents regardless of where it is produced, whereas GDP measures total production within a country regardless of who owns the means of production. Fishman and Fishman, *Employment*, 2; Kennedy, *Freedom from Fear*, 619.

2. Senate Temporary National Economic Committee, *Investigation of Concentration*, April 26, 1940, 17238–239; Hart-Landsberg, "Popular Mobilization," 401–2. See data on unemployment and GDP, Federal Reserve Economic Data, Federal Reserve

Bank of St. Louis, accessed September 7, 2020, https://fred.stlouisfed.org/series/A191RP1A027NBEA.

3. Dallek, *Franklin D. Roosevelt*, 385–402.

4. Heath, "American War Mobilization," 297–99; Hart-Landsberg, "Popular Mobilization," 402; Dallek, *Franklin D. Roosevelt*, 402–16.

5. "Number of Jobless Persons Decline," *Wall Street Journal*, May 20, 1942; Frank Woodford, "Jobs at Detroit at Record High," *New York Times*, June 14, 1942; Franklin D. Roosevelt, State of the Union address, January 6, 1942, American Presidency Project, https://www.presidency.ucsb.edu/documents/state-the-union-address-1. Doris Kearns Goodwin quotes the columnist Raymond Clapper; see Goodwin, *No Ordinary Time*, 482.

6. Lichtenstein, *Walter Reuther*, 337–38. Vatter, *U.S. Economy*, 12–13; Lichtenstein, *Contest of Ideas*, 82–83.

7. Miller, "Learning from History"; Dechter and Elder, "World War II Mobilization"; Field, "Impact of the Second World War," 681–82.

8. Vatter, *U.S. Economy*, 17; "Technology for War," *Wall Street Journal,* April 5, 1945.

9. "Number of Jobless Persons Decline"; Woodford, "Jobs at Detroit"; "Remarks [by] Senator Harvey Kilgore before Subcommittee of the Committee on Military Affairs," *Congressional Record*, 77th Cong., 2nd sess., October 13, 1942, vol. 88, pt. 6, 8107–8; Goodwin, *No Ordinary Time*, 625; "American Scientists Are Mobilized for War," *Washington Post*, May 17, 1942.

10. Wiener, *Cybernetics*, 1–29; Conway and Siegelman, *Dark Hero*, 48–55, 63–67, 138–43.

11. Zachary, *Endless Frontier*, 23–56.

12. Zachary, *Endless Frontier*, 48–52.

13. Zachary, *Endless Frontier*, 75–83; Owens, "Counterproductive Management."

14. Zachary, *Endless Frontier*, 75–83; Owens, "Counterproductive Management"; Goldberg. "Inventing a Climate."

15. Baxter, *Scientists against Time*, vii–viii, 131–37, 234–37.

16. Conway and Siegelman, *Dark Hero*, 104, 111–18. Wiener engaged in a vital correspondence with Bush after he left MIT and would continue to write him during the war.

17. "Defense Strikes Assailed at Tufts," *New York Times*, June 16, 1941.

18. Kennedy, *Freedom from Fear*, 296–98; Dray, *Power in a Union*, 485–91. The Bureau of Labor Statistics counted the number of union employees after World War II at an estimated fifteen million; *Directory of Labor Unions*, 1. At that point the total workforce was approximately sixty million individuals; see Federal Reserve Economic Data, "Total Civilian Labor Force (16 Years and Older) for United States," Federal Reserve Bank of St. Louis, accessed September 23, 2020, https://fred.stlouisfed.org/series/M0825BUSM148NNBR; Lambert, "*Workers Took a Notion*," 105–28.

19. Dray, *Power in a Union*, 485–91; Lambert, "*Workers Took a Notion*," 105–28.

20. Dray, *Power in a Union*, 485–91; Wall, *Inventing the "American Way*," 193; Lichtenstein, *State of the Union*, 100; Lichtenstein, *Labor's War at Home*, 203–6; Lambert, "*Workers Took a Notion*," 105–28.

21. Lichtenstein, *Labor's War at Home*, 203–6 (Nelson quote appears on 205); Wall, *Inventing the "American Way*," 167–69; Jones, "Freedom from Want"; Web, "Natural Rights."

22. Wilkerson, *Warmth of Other Suns*, 216–17; Wall, *Inventing the "American Way,"* 153–54.

23. Jacobs, "Uncertain Future," 155; Lichtenstein, *State of the Union*, 216; Noble, *Forces of Production*, 45–56.

24. Hansen, *After the War*, 1–16. According to Alan Brinkley, the NRPB's plan "was a product . . . of a firm commitment to the goal of a 'high-income, full-employment' economy and of a belief that programs of public aid would help create it"; Brinkley, *End of Reform*, 252. Also see "The National Resources Planning Board," Executive Order No. 6777, June 20, 1934, Public Papers and Addresses of Franklin D. Roosevelt, Internet Archive, https://archive.org/details/4925383.1934.001.umich. edu/page/335/mode/2up.

25. Brinkley, *End of Reform*, 265–66; Pursell, *Technology in Postwar America*, 78–81; Hounshell, *American System*, 217–61; "The Automatic Factory," *Fortune*, November 1946, 160; Nye, *America's Assembly Line*, loc. 2920–85 of 7470, Kindle; Dobney, "Reconversion Policy"; Goodwin, *No Ordinary Time*, 315.

26. "30-Hour Week Plan by Reuther Rouses Johnston, Kaiser," *New York Times*, March 17, 1944; *Encyclopedia Britannica Online*, s.v. "Henry J. Kaiser," accessed December 17, 2019, https://www.britannica.com/biography/Henry-J-Kaiser.

27. Dobney, "Reconversion Policy," 498–510 (references to *Newsweek* and the *Saturday Evening Post* appear on p. 510, notes 52 and 53, and the reference to *Fortune* is on p. 501, note 12); Ben W. Gilbert, "Baruch Calls for Start Now of 'X-Day' Reconversion," *Washington Post*, February 19, 1944.

28. "Labor Force Statistics from the Current Population Survey," US Bureau of Labor Statistics, accessed September 5, 2020, https://www.bls.gov/cps/aa2014/cpsaat01.htm. Roosevelt's prophecy came true in 1947 when employment in the United States reached just over sixty million. See also "Employment Status of Persons on the Labor Force, 1940," *Statistical Abstract*, 123, table 131. The Roosevelt quote appears in the testimony of George E. Outland, Senate Subcommittee on Full Employment, *Full Employment Act*, July 31, 1945, 128.

29. Brinkley, *End of Reform*, 226–45. "Report of the Director of War Mobilization Activities under Executive Order 9425," October 31, 1944, notes that the executive order followed by four days the submission of the Baruch-Hancock report; see "Report to the Director of War Mobilization as to Activities Under Executive Order 9425," accessed October 12, 2019, https://babel.hathitrust.org/cgi/pt?id=nnc1.cu04159713&view=1up&seq=3.

30. Baruch and Hancock, *Report on War*, 1–3; Dobney, "Reconversion Policy"; "War Mobilization and Post-war Adjustment," *Congressional Record*, 78th Cong., 2nd sess., March 29, 1944, vol. 90, pt. 3, 3241; Grant, *Bernard Baruch*, 273–89; Dobney, "Reconversion Policy," 498–504; Brinkley, *End of Reform*, 244.

31. Brinkley, *End of Reform*, 244–45; Key, "Reconversion Phase"; Dobney, "Reconversion Policy," 498–504.

32. Dobney, "Reconversion Policy," 498–504; Ben W. Gilbert, "Baruch Spurs WPB, Congress on Conversion," *Washington Post*, June 15, 1944; "Baruch Out of Postwar Unit; Reconversion Plan Stalled," *Washington Post*, June 7, 1944; Joint Special Committee on Postwar Economic Policy and Planning, *Postwar Economic Policy*, June 16, 1944, 1–79; Koistinen, "Mobilizing the World War II Economy," 446; Senate Committee on Post-war Economic Policy and Planning, *Report on Post-war Economic Policy*, 5–7; Brinkley, *End of Reform*, 241–46.

33. Leuchtenburg, *White House Looks South*, 91, 449; "Roosevelt Asks Defeat of George and Talmadge as Foes of Liberalism," *New York Times*, August 12, 1938.

34. Leuchtenburg, *White House Looks South*, 91, 449; Wasem, *Tackling Unemployment*, loc. 682 of 5864, Kindle; Dobney, "Reconversion Policy," 502–3; Senate Committee on Post-war Economic Policy and Planning, *Report on Post-war Economic Policy*, 5–7; Brinkley, *End of Reform*, 241–46.

35. Senate Subcommittee on Full Employment, *Full Employment Act*, August 28, 1945, 507–15.

36. Joint Special Committee on Postwar Economic Policy and Planning, *Postwar Economic Policy and Planning*, July 27, 1944, 1709–24; House Special Committee on Postwar Economic Policy and Planning, *Postwar Economic Policy and Planning*, 79th Cong., 1st sess., March 13, 1945, 1846; Smith, *Building New Deal Liberalism*, 234–48; Wasem, *Tackling Unemployment*, loc. 658 of 5864, Kindle.

37. James E. Murray, "Contract Settlement Act of 1944," Duke Law Scholarship Repository, accessed February 12, 2020, https://scholarship.law.duke.edu/cgi/viewcontent.cgi?article=2210&context=lcp; Cain and Neumann, "Planning for Peace."

4. Full Employment

1. Senate Subcommittee on Full Employment, *Full Employment Act*, July 30, 1945, 1–5; "Unknown Number Missing, Others Are Dying," *Washington Post*, August 23, 1945; "We Enter a New Era—the Atomic Age," *New York Times*, August 12, 1945.

2. Harold Fleming, "Story of New 'Dream' Goods, Gift of Technology, Now Told," *Christian Science Monitor*, September 7, 1945; Dorothy Thompson, "The Struggle Is Not Over," On the Record, *Atlanta Constitution*, August 15, 1945; "Sees Record Productivity after Five Years," *Christian Science Monitor*, December 26, 1945; Harold Fleming, "Basic Research Drags While Applied Generates New Goods," *Christian Science Monitor*, September 11, 1945.

3. Beardsley, *1945 Collier's Yearbook*, iii, 43, 422, 455, 464, 476; "U.S. Promised Era of Gadgets—If Industrial Strife Is Checked," *Christian Science Monitor*, November 10, 1945.

4. Phillips-Fein, *Invisible Hands*, 112–18; Waterhouse, *Lobbying America*, 18–19. Waterhouse suggests that the National Association of Manufacturers and the Chamber of Commerce increasingly were viewed as out of step: "Indeed, their unflinching faith in self-correcting markets and staunch opposition to Keynesianism combined to minimize their influence." Mayhew, "Long 1950s," 26–29, makes the case that Wilson, as well as other more progressive minds in the business community, soon found a home in the newly formed Committee on Economic Development. Also see Cohen, *Consumer's Republic*, 114–19.

5. A comprehensive discussion of the evolution of the term "liberalism" as it evolved after the Wilsonian progressive era can be found in Mileur, "Great Society," 418–20; Young, "'Do Something,'" 202–3.

6. "Reuther Challenges Our Fear of Abundance," *New York Times*, September 16, 1945.

7. Wall, *Inventing the "American Way,"* 167–73; US Bureau of Labor Statistics, *Work Stoppages Caused by Labor-Management Disputes in 1945*, 1–5; Cohen, *Consumer's Republic*, 69; "Report of the Advisory Committee," 16–22; US Bureau of Labor

Statistics, *Work Stoppages Caused by Labor-Management Disputes in 1946*, 1, https://fra
ser.stlouisfed.org/files/docs/publications/bls/bls_0918_1947.pdf; "750,000 Steelworkers on Strike," *Los Angeles Times*, January 21, 1946; Phillips-Fein, *Invisible Hands*,
115–16.

8. "President Shuns Mine Seizure," *Christian Science Monitor*, April 1, 1946; Dray,
Power in a Union, 491–502; "Truman's Bold Action Changes the Picture," *New York
Times*, May 26, 1946.

9. Wall, *Inventing the "American Way,"* 167–73; Wilkerson, *Warmth of Other Suns*,
19–27; Keri Pleasant, "Honoring Black History World War II Service to the Nation,"
US Army, February 27, 2020, https://www.army.mil/article/233117/honoring_
black_history_world_war_ii_service_to_the_nation; Young, "'Do Something,'"
199–201.

10. Franklin D. Roosevelt, State of the Union address, January 6, 1941, University of Virginia Miller Center, https://millercenter.org/the-presidency/presidential-
speeches/january-6-1941-state-union-four-freedoms (this was the State of the Union
during which Roosevelt articulated the "four freedoms"); Lippmann, "Discovered in
Our Time."

11. Beveridge, *Full Employment*, 18; Bailey, *Congress Makes a Law*, 228.

12. Beveridge, *Full Employment*, 18–21; Haber, "Strategy of Reconversion"; Harold Fleming, "Full Employment Goal Seen as Doom of Balanced Budget," *Christian
Science Monitor*, October 23, 1944.

13. Senate Subcommittee on Full Employment. *Full Employment Act*, July 30,
1945, 1–5.

14. S. 380, introduced by James Murray (D-MT) in the Senate on January 22, 1945,
was cosponsored by Robert Wagner (D-NY), Joseph O' Mahoney (D-WY), Elbert
Thomas (D-UT), Wayne Morse (R-OR), George Aiken (R-VT), William Langer
(R-ND), and Charles Tobey (R-NH). Section 3 of the bill details the requirements of
the National Production and Employment Rights Budget. Senate Subcommittee on
Full Employment, *Full Employment Act*, July 30, 1945, 1–18.

15. Wasem, *Tackling Unemployment*, loc. 1472–1528 of 5864, Kindle; De Long,
"Keynesianism."

16. De Long, "Keynesianism." For an analysis of the demise of the National
Resources Planning Board and the fault lines of the Full Employment Act debate,
see also Brinkley, *End of Reform*, 259–61. In addition to his role as head of the National
Farmers Union, Patton was a trustee of the National Planning Association, a diverse
group of farm, labor, and business leaders that had largely supported New Deal
stimulus and was at the center of support for full employment legislation. Bailey,
Congress Makes a Law, 20–28; Livermore, "James G. Patton," 107–15; US Department
of Agriculture, *Technology on the Farm*, 61; House Committee on Expenditures, *Full
Employment Act*, October 25, 1945; Senate Subcommittee on Full Employment, *Full
Employment Act*, August 24, 1945, 319–29; Interbureau Committee, "Revolution on
the Farm"; Dimitri, Effland, and Conklin, *20th Century Transformation*, 2, 6–8; Arthur
J. Goldberg, "The Challenge of 'Industrial Revolution II,'" *New York Times*, April 2,
1961.

17. Senate Subcommittee on Full Employment, *Full Employment Act*, July 30,
1945, 223–37.

18. Young, "'Do Something,'" 199–201. As Alan Brinkley notes, "The G.I. Bill provided veterans with many of the benefits the NRPB, and other liberals, had hoped to provide all Americans after the war." Brinkley, *End of Reform*, 258.

19. Paul A. Samuelson, "Full Employment," *Washington Post*, September 2, 1945.

20. Jacobs, "Uncertain Future," 154–55; Bailey, *Congress Makes a Law*, 150–53; "Wright Patman, Dean of House, Dies," *New York Times*, March 8, 1976; House Committee on Expenditures, *Full Employment Act*, September 25, 1945, 38.

21. Dorothy Thompson, "The Struggle Is Not Over," *Atlanta Constitution*, August 15, 1945. Outland was a cosponsor of Patman's H.R. 2202 in the House, and led a "Dear Colleague" letter generating 103 signatures. Wasem, *Tackling Unemployment*, loc. 1957–2420 of 5864, Kindle.

22. Wasem, *Tackling Unemployment*, loc. 1957–2420 of 5864, Kindle; House Committee on Expenditures, *Full Employment Act*, October 25, 1945, 802.

23. Hoffman was known to be irritable. *Time* magazine would subsequently describe him as "a bitter lone wolf . . . perhaps the most reactionary man in Congress"; see "National Affairs: Old Faces," *Time*, November 17, 1952; House Committee on Expenditures, *Full Employment Act*, September 25, 1945, 35–49, and October 25, 1945, 347–91; Bailey, *Congress Makes a Law*, 249–52, lists cosponsors of H.R. 2022.

24. "Full Employment Act of 1945," *Congressional Record*, 79th Cong., 1st sess., September 28, 1945, vol. 91, pt. 7, 9123–53; Bailey, *Congress Makes a Law*, 104–27; "Record Vote in Senate on Full Employment," *New York Times*, September 29, 1945.

25. Wasem, *Tackling Unemployment*, loc. 1746–853 of 5864, Kindle; "Full Employment Act of 1945," *Congressional Record*, 79th Cong., 1st sess, September 27, 1945, vol. 91, pt. 7, 9047–92.

26. Bailey, *Congress Makes a Law*, 164–73; Wasem, *Tackling Unemployment*, loc. 2318–420 of 5864, Kindle.

27. Harry S. Truman, "Special Message to the Congress Presenting a 21-Point Program for the Reconversion Period," September 6, 1945, Harry S. Truman Library and Museum, https://www.trumanlibrary.gov/library/public-papers/128/special-message-congress-presenting-21-point-program-reconversion-period; Harry S. Truman, "Letter to Senator Wagner and Representative Manasco concerning the Full Employment Bill," December 20, 1945, Harry S. Truman Library and Museum, https://www.trumanlibrary.gov/library/public-papers/222/letter-senator-wagner-and-representative-manasco-concerning-full; "Employment Production Act," *Congressional Record*, 79th Cong., 1st sess., December 14, 1945, vol. 91, pt. 9, 12063–65; Bailey, *Congress Makes a Law*, 150–77; Wasem, *Tackling Unemployment*, loc. 2311–417 of 5864, Kindle.

28. Bailey, *Congress Makes a Law*, 222–27; Employment Act of 1946, PL 79-304, 79th Cong., 2nd sess., February 20, 1946. Sensing defeat, Wagner relied on Bertram Gross, a key staff member on the Banking Committee and a critical drafter of the original legislation, as a surrogate during the conference, where conservatives outnumbered liberal-labor conferees. "Bertram M. Gross, 84, Author of Full Employment Bills," *New York Times*, March 15, 1997.

29. "Employment Production Bill-Conference Report (S. 380)," *Congressional Record*, 79th Cong., 2nd sess., February 8, 1946, vol. 92, pt. 1, 1139. The discussion

of the Roper poll appears in Wasem, *Tackling Unemployment*, loc. 1272–99 of 5864, Kindle. The discussion of the Senate debate appears in Wasem, *Tackling Unemployment*, loc. 1743– 853 of 5864, Kindle. "Statement by President Harry S. Truman on Signing the Full Employment Act of 1946," February 20, 1946, Harry S. Truman Library and Museum, https://www.trumanlibrary.gov/library/public-papers/39/statement-president-upon-signing-employment-act; "Employment Bill Signed by Truman," *New York Times*, February 21, 1946; "Bertram M. Gross."

30. "Employment Production Bill," 976, 981. Manasco may have understated the national debt. Indeed, because of the war mobilization, the government was $279 billion in debt, which amounted to 112 percent of GDP. Matt Phillipps, "The Long Story of U.S. Debt, from 1790 to 2011, in 1 Little Chart," *Atlantic*, November 13, 2012; "House Votes Job Bill—but Not '60 Million' Kind," *Chicago Daily Tribune*, February 7, 1946; "Employment Bill—1946," *Washington Post*, February 9, 1946.

31. Brinkley, *End of Reform*, 261; Aaron Steelman, "Employment Act of 1946," Federal Reserve History, November 22, 2013, https://www.federalreservehistory.org/essays/employment_act_of_1946; "Franklin Roosevelt Call for Federal Responsibility," excerpt from October 13, 1932, campaign speech, Columbia University, http://www.columbia.edu/~gjw10/fdr.newdeal.html (also cited in Milkis, "Franklin Roosevelt," 39); "Floor Debate of 'Full Employment Production Bill' Conference Report," *Congressional Record*, 79th Cong., 2nd sess., February 6, 1946, vol. 92, pt. 1, 975–82.

32. "No Politics, Nourse Says," *Washington Post*, July 31, 1946; "Nourse Is Named Truman Adviser," *Baltimore Sun*, July 30, 1946; "Remarks by Senator Wagner on Confirmation of Edwin G. Nourse to Be Chairman, National Council of Economic Advisors," *Congressional Record*, 79th Cong., 2nd sess., July 30, 1946, vol. 92, pt. 8, 10459; Senate Committee on Military Affairs, *Mobilization and Demobilization Problems*, April 26, 1944, 90–103; Knapp, *Edwin G. Nourse*, 200–201, 226–28.

33. Council of Economic Advisers, *Economic Report of the President*, 1954, 64, 162; Pursell, *Technology in Postwar America*, 79–81; "Precision Devices in Better Supply," *New York Times*, February 14, 1948.

34. Leaver moved to Canada as a child, and soon after graduating from high school he began tinkering with new landing systems for aircraft. During the war he served in Canada's wartime radar research office, and after the war he formed his own firm aimed at developing automated machine tools. "Eric William Leaver," *Canada Historica*, last updated December 22, 2017, http://www.thecanadianencyclopedia.ca/en/article/eric-william-leaver/; "Arthur Lidov, 73, Artist and Inventor," *New York Times*, January 2, 1981; "A Note on the Authors," *Fortune*, November 1946, 196; E. W. Leaver and J. J. Brown, "Machines without Men," *Fortune*, November 1946, 165, 192–96; "The Automatic Factory," *Fortune*, November 1946, 160–64.

35. Noble, *Forces of Production*, 57–61.

36. Council of Economic Advisers, *Economic Report of the President*, 1946, 2–32; Brinkley, *End of Reform*, 260–67.

37. Council of Economic Advisers, *Economic Report of the President*, 1947, vii.

38. Reference to the pamphlet *It's Fun to Live in America* in Pursell, *Technology in Postwar America*, 97–98; Bush, "Science," 232.

5. Automation Arrives

1. Lichtenstein, *Walter Reuther*, 260–70; Conway and Siegelman, *Dark Hero*, 237–39; Wiener, *Cybernetics*, 27–29, 36–37. The text of the exchange between Wiener and Reuther appears in Noble, *Progress without People*, loc. 3153–201 of 3229, Kindle.

2. "Labor: The G.A.W. Man," *Time*, June 20, 1955; Noble, *Progress without People*, loc. 3153–201 of 3229, Kindle. Membership numbers for the UAW and other unions provided in Nelson, "How the UAW Grew," 11; Lichtenstein, *Walter Reuther*, 291–93; "Ford Workers Vote for Strike by 7–1 Margin," *Chicago Tribune*, August 12, 1949.

3. Lichtenstein, *Walter Reuther*, 291–93.

4. "Auto Union Warns General Motors," *New York Times*, May 8, 1950; Lichtenstein, *Walter Reuther*, 278–81 (Lichtenstein notes that Bell and Reuther became acquainted while they were members of the Socialist Party during the 1930s); "G.M. Workers Reach 5-Year Pact on Pensions, Wages," *New York Times*, May 24, 1950; "Guaranteed Annual Pay Next Goal for Reuther," *Christian Science Monitor*, May 29, 1950.

5. Conway and Siegelman, *Dark Hero*, 246–54.

6. "Automation Guides Auto Valve Bushings," *Christian Science Monitor*, December 1948; "Ford Adds Three Plants Facilities to Cleveland Operations," *Wall Street Journal*, December 26, 1951; "Cleveland," *New York Times*, April 26, 1953; Walter Reuther, "Automation Compounds Maintenance Problems," *Mill and Factory*, October 5, 1953, 93; Meyer, "Economic Frankenstein"; Lichtenstein, *Walter Reuther*, 290–92; see also Walter Reuther testimony (roughly five minutes from the beginning) in QuestexMediaGroup, "Test Video—Push Buttons and People," YouTube video, 20:25, June 13, 2013, https://www.youtube.com/watch?v=GR9kVCoJAr8.

7. Dwight D. Eisenhower, "Annual Message to the Congress on the State of the Union," February 2, 1953, Dwight D. Eisenhower Presidential Library, Museum and Boyhood Home, 15, https://www.eisenhowerlibrary.gov/sites/default/files/file/1953_state_of_the_union.pdf. Federal research spending as a percentage of GDP reached its highest levels during the Eisenhower administration and has not been exceeded since; see Usselman, "Research and Development," 21–23; Newton, *Eisenhower*, locs. 207, 1353, of 9007, Kindle; Bird and Sherwin, *American Prometheus*, loc. 6238 of 15955, Kindle; Polsby, *Political Innovation*, 16–17.

8. Vannevar Bush. "Science: The Endless Frontier," July 1945, National Science Foundation, https://www.nsf.gov/od/lpa/nsf50/vbush1945.htm; England, *Patron for Pure Science*, 5–10; Newton, *Eisenhower*, locs. 207, 1353, of 9007, Kindle; Bird and Sherwin, *American Prometheus*, loc. 6238 of 15955, Kindle; Polsby, *Political Innovation*, 16–17.

9. For a detailed analysis of the political machinations over the formation of the NSF from the 79th to the 81st Congresses, see England, *Patron for Pure Science*, 9–106; Wang, *American Science*, 25–34; Polsby, *Political Innovation*, 35–55; Zachary, *Endless Frontier*, 325–26, 342–43.

10. Shermer, *Sunbelt Capitalism*, 4.

11. Shermer, *Sunbelt Capitalism*, 132–41.

12. Shermer, *Sunbelt Capitalism*, 38–45.

13. "Fair Deal Economic Expansion Bill Passage This Year Held Unlikely," *Wall Street Journal*, July 16, 1949; "Senate Bill to Implement Truman Fight on Recession," *New York Times*, July 16, 1949; "Remarks on the Economic Expansion Bill of 1949," *Congressional Record*, 81st Cong., 1st sess., July 15, 1949, vol. 95, pt. 7, 9537–43; Wilson, *Communities Left Behind*, xii–xvi, 22–23; Newton, *Eisenhower*, locs. 207, 1353, of 9007, Kindle.

14. Shaw, *JFK in the Senate*, 19–29, 52, 58; the quote from Roosevelt is found in Caro, *Path to Power*, 619; the quote from Smathers is found in Dallek, *Lyndon B. Johnson*, 83; Caro, *Means of Ascent*, 758.

15. Caro, *Master of the Senate*, 439–51.

16. Caro, *Master of the Senate*, 472–75; Shaw, *JFK in the Senate*, 50–52.

17. Shaw, *JFK in the Senate*, 53, 70; Oliphant and Wilkie, *Road to Camelot*, 84–86, 105.

18. John Harris, "Kennedy Offers Ten Point Plan for New England," *Boston Globe*, May 19, 1953.

19. Dallek, *Lyndon B. Johnson*, 181; "Remarks of Senator Kennedy on New England Economy," *Congressional Record*, 83rd Cong., 1st sess., May 18, 1953, May 20, 1953, May 25, 1953, vol. 99, pt. 4, 5054–64, 5227–40, 5455–66.

20. "Remarks of Senator Kennedy," May 18, 1953, May 20, 1953, May 25, 1953, 5054–72, 5227–40, 5455–66; John F. Kennedy, "New England and the South," *Atlantic Monthly*, January 1954, https://www.theatlantic.com/magazine/archive/1954/01/new-england-and-the-south/376244/.

21. "Remarks of Senator Kennedy," May 18, 1953, May 20, 1953, May 25, 1953, 5054–72, 5227–40, 5455–66; Kennedy, "New England."

22. "Remarks of Senator Kennedy," May 18, 1953, 5058; Bachmura, "Manpower Development"; Gladys Roth Kremen, "The Origins of the Manpower Development and Training Act," US Department of Labor, accessed April 24, 2021, https://www.dol.gov/general/aboutdol/history/mono-mdtatext. The Area Redevelopment Act of 1961 resulted in the investment of some $400 million into the private sector to stimulate new job creation, particularly in depressed areas. Some $4 million was dedicated to job training.

23. Gross domestic product (then measured as the gross national product; see chapter 3, note 1) began declining in the first quarter of 1953, but a year later had begun its recovery, and by the first quarter of 1955 was growing 14 percent faster than in the first quarter of 1954. See Federal Reserve Economic Data, "Gross National Product," Federal Reserve Bank of St. Louis, accessed April 24, 2021, https://fred.stlouisfed.org/series/A001RP1Q027SBEA; Council of Economic Advisers, *Economic Report of the President*, 1954, iv, 20; Council of Economic Advisers, *Economic Report of the President*, 1955, 29, 54–57, 88–92; "Program to Alleviate Conditions of Excessive Unemployment in Certain Areas," *Congressional Record*, 84th Cong., 1st sess., July 28, 1955, vol. 101, pt. 9, 11754–57. The minimum wage was raised from forty cents an hour to seventy-five cents an hour in 1949, and a 1955 amendment sponsored by Humphrey then raised it to one dollar an hour. See "History of Changes to the Minimum Wage Law," US Department of Labor, Wage and Hour Division, accessed May 18, 2020, https://www.dol.gov/agencies/whd/minimum-wage/history.

24. Joint Committee on the Economic Report, *January 1954 Economic Report*, 772–75; Reuther, "Policies for Automation"; Stanley Ruttenberg, "Prediction on Unemployment," *New York Times*, December 17, 1954.

25. "Reuther Asks for U.S. Job Aid in Rise of Automated Plants," *New York Times*, December 6, 1954; "Guaranteed Yearly Wage Set-Up as Goal for CIO," *Los Angeles Times*, December 7, 1954; Joint Committee on the Economic Report, *Hearings on Automation*, October 17, 1955, 98–105; "Guaranteed Annual Wage," Congressional Quarterly Researcher, January 21, 1953, https://library.cqpress.com/cqresearcher/document.php?id=cqresrre1953012100; Waldemar Kaempffert, "Walter Reuther's Fears of Automation Stir New Interest in Factory of the Future," *New York Times*, December 12, 1954.

26. "Labor: The G.A.W. Man," *Time*, June 20, 1955; "What about GAW," *Christian Science Monitor*, April 8, 1955; House Committee on Expenditures in the Executive Departments. *Full Employment Act of 1945*, October 16, 1945; Shaffer, "Guaranteed Annual Wage"; Kaplan, *Guarantee of Annual Wages*, 1–12, 63–64, 213–15; "Guaranteed Annual Wage"; Bernstein, *Lean Years*, 485, 503–4; Dray, *Power in a Union*, 458–59; US House of Representatives, *Supplemental Estimate of Appropriations*; "Annual Wage Plan Pushed by Truman Urging More Study," *New York Times*, March 9, 1947; "S. 889, Senator McMahon," *Congressional Record*, 80th Cong., 1st sess., March 14, 1947, vol. 93, pt. 2, 2054.

27. "Guaranteed Yearly Wage Set-Up"; "Opening the Drive for a Guaranteed Annual Wage," *Baltimore Sun*, March 6, 1955; "Guaranteed Wage Victories," *Washington Post*, August 17, 1955; "GM Signs Guaranteed Wage Pact," *Washington Post*, June 14, 1955; "What about GAW"; Joint Committee on the Economic Report, *Hearings on Automation*, October 17, 1955, 98–105, 128; "Guaranteed Annual Wage"; Kaempffert, "Walter Reuther's Fears"; Lichtenstein, *Walter Reuther*, 284–86; "Ford and Union Reach 3-Year Pact including Modified Annual Wage," *New York Times*, June 7, 1955; "Union Threatens Strike Unless General Motors Betters Ford Terms," *Wall Street Journal*, June 8, 1955; "Guaranteed Wage Due for Showdown," *New York Times*, March 23, 1947; Latimer, *Guaranteed Wages Report*, 164–69; "Administration Study Urges Firms to Set Guarantee Wages; Unions Favor It; Want Higher Wages First," *Wall Street Journal*, March 19, 1947.

28. Joint Committee on the Economic Report, *Report of the Joint Committee*, 23–24; Joint Committee on the Economic Report, *Hearings on Automation*, October 14, 1955, 3. The six industries that were the focus of the Patman automation hearings were metalworking, data processing, chemicals, electronics, transportation, and communications.

29. Joint Committee on the Economic Report, *Hearings on Automation*, October 17, 1955, 97–103. Reuther cited a number of examples of automation's incursion into the workplace. He noted that a concrete supply company in Cleveland could produce one of 1,500 different mixes and load it onto trucks without any manual labor; a single automatic machine at the Cincinnati Milling Machine Company produced the same volume of cylinder heads previously produced by 162 separate machines and cut the company's costs by 80 percent; Honeywell in Minneapolis was using an electronically controlled boring machine to bring precision levels to an infinitesimal one-thousandth of an inch; and the Bank of America's "electronic brains" were implementing a new check processing machine that eliminated vast numbers of clerical employees and was capable of storing information on some thirty-two thousand accounts.

30. Joint Committee on the Economic Report, *Hearings on Automation*, October, 17, 1955, 97–114, 246–49, and October 28, 1955, 618–27. The quote from the NAM pamphlet appears on p. 102.

31. "Detroit Is Leading the Trend toward Full Mechanization," *New York Times*, February 14, 1954; Joint Committee on the Economic Report, *Hearings on Automation*, October 14, 1955, 1–2; *Economic Report of the President*, January 20, 1955, iii–iv, 23, 51, 56, 88–91.

32. "James P. Mitchell Is Dead at 63," *New York Times*, October 20, 1964; Joint Committee on the Economic Report, *Hearings on Automation*, October 24, 1955, 262–91.

33. Joint Committee on the Economic Report, *Hearings on Automation*, October 14, 1955, 6–48; Jennifer Bayott, "John Diebold, 79, a Visionary of the Computer Age, Dies," *New York Times*, December 27, 2005; Drucker, foreword to *Beyond Automation*, viii; Slesinger, "Pace of Automation," 241; Valerie J. Nelson, "John Diebold, 79; Pioneered Computer Use in Automation of Businesses," *Los Angeles Times*, December 30, 2005; "Automation: Industrialists Call for Careful Definition," *Christian Science Monitor*, October 19, 1955; Diebold, "Automation"; Cross, *John Diebold*, 58–59.

34. Joint Committee on the Economic Report, *Hearings on Automation*, October 28, 1955, 603–18; Zachary, *Endless Frontier*, 385–89.

35. Council of Economic Advisers, *Economic Report of the President*, 1956, iii, 7, 12, 51, 87; Joint Committee on the Economic Report, *Automation and Technological Change*, 4–14.

36. Wilson, "Before the Great Society," 121–22; "Trade Adjustment Act of 1955," *Congressional Record*, 84th Cong., 1st sess., May 4, 1955, vol. 101, pt. 4, 5566, 5572.

37. "G. Mennen Williams, Ex-Michigan Governor," *Chicago Tribune*, February 3, 1988; "A Program to Alleviate Unemployment Conditions in Certain Economically Depressed Areas," *Congressional Record*, 84th Cong., 2nd sess., June 11, 1956, vol. 102, pt. 7, 10011–15; Joint Committee on the Economic Report, *Automation and Technological Change*, 2–13; Senate Labor Subcommittee on Labor and Public Welfare, *Area Redevelopment Hearings*, January 23, 1956, 88–90, 229–55.

38. "Daniel Flood, 90, Who Quit Congress in Disgrace, Is Dead," *New York Times*, May 29, 1994; Kashatus, *Dapper Dan*, 97–102; House Committee on Banking and Currency, *Legislation to Relieve Unemployment*, May 20, 1958, 1176–78; Senator Paul H. Douglas, "Program to Alleviate Conditions of Excessive Unemployment in Certain Areas," *Congressional Record*, 84th Cong., 1st sess., July 28, 1955, vol. 101, pt. 9, 11754–57; Biles, *Crusading Liberal*, 150–55; Douglas, *Fullness of Time*, 512–22.

39. House Committee on Banking and Currency, *Legislation to Relieve Unemployment*, April 14, 15, 21, 22, 1955, pp. 69, 122, 132, 263, 312.

40. "Area Redevelopment Bill Vetoed."

41. Senate Special Committee on Unemployment Problems, *Hearings on Unemployment Problems*, 1–4; Senate Special Committee on Unemployment, *Report of the Special Committee*, 1–15.

42. Senate Special Committee on Unemployment, *Report of the Special Committee*, 47; Senate Special Committee on Unemployment, *Hearings before the Special Committee, Part 8*. See Senator Gale McGee's statement on extractive industries in Wyoming, iii–iv; and statement of Samuel I. Brooks, South Bend Office Indiana Employment Section, 3116–27.

43. Senate Special Committee on Unemployment, *Hearings on Unemployment Problems, Part 6*, 2510–12; remarks of Congressman John Slack, "Gone-Employment Is the Real Menace," *Congressional Record*, 86th Cong., 1st sess., June 3, 1959, vol. 105, pt. 7, 9748–50.

44. Senate Special Committee on Unemployment, *Report of the Special Committee*, 44–48, 120–26, 133–35. The report notes on p. 122, "Class rather than mass unemployment has characterized the past decade."

45. Rick Perlstein describes Senator Hugh Scott as "a liberal Republican," a broad term encompassing mostly eastern-state Republicans who were fiscally conservative and "hawkish on defense," but more socially progressive; see Perlstein, *Nixonland*, 80. Arthur Schlesinger describes Kennedy as speaking "with enthusiasm" about Kentucky's Republican senator John Sherman Cooper; see Schlesinger, *Thousand Days*, 18. Also see Senate Special Committee on Unemployment, *Report of the Special Committee*, 179–89.

46. "Area Redevelopment Act," *Congressional Record*, 86th Cong., 2nd sess., May 4, 1960, vol. 106, pt. 7, 9417–72; Biles, *Crusading Liberal*, 150–55; Douglas, *Fullness of Time*, 512–22.

6. Creating a Commission

1. "John F. Kennedy Statement Announcing Candidacy for President," January 2, 1960, John F. Kennedy Presidential Library and Museum (hereafter Kennedy Library), https://www.jfklibrary.org/asset-viewer/archives/JFKSEN/0905/JFKSEN-0905-021.

2. White, *Making of the President 1960*, 95–125.

3. Harrison Salisbury, "West Virginia: Battleground for Democrats," *New York Times*, May 2, 1960; Sorensen, *Kennedy*, 402; Philip Benjamin, "Kennedy Gloomy on West Virginia," *New York Times*, May 6, 1960; "Kennedy Gigs in Coal Fields," *Christian Science Monitor*, April 26, 1960.

4. "Kennedy Charges National Decline, Williams Endorses Him," *New York Times*, June 3, 1960; Sorensen, *Kennedy*, 148–49; Warren, "Politics of Labor Policy Reform."

5. Reuther, "Policies for Automation"; Lichtenstein, *Walter Reuther*, 291.

6. John F. Kennedy, "Labor: Meeting the Problems of Automation," AFL-CIO convention, Grand Rapids, MI, June 7, 1960, Kennedy Library, https://www.jfklibrary.org/Asset-Viewer/Archives/JFKCAMP1960-1030-036.aspx; "Automation Is Tackled by Kennedy," *Baltimore Sun*, June 8, 1960.

7. Kennedy, "Labor"; "Kennedy a Hit with Labor," *Detroit Free Press*, June 8, 1960; "Remarks of Senator John F. Kennedy at Springfield College, Springfield, Massachusetts," June 10, 1956, Kennedy Library, https://www.jfklibrary.org/archives/other-resources/john-f-kennedy-speeches/springfield-ma-19560610.

8. Dallek, *Unfinished Life*, 287; John F. Kennedy, "Weyerhaeuser Lumber Company, Eugene, Oregon," May 17, 1960, Kennedy Library, https://www.jfklibrary.org/asset-viewer/archives/JFKSEN/0909/JFKSEN-0909-034; "Remarks of Senator John F. Kennedy, Pocatello, Idaho," September 6, 1960, Kennedy Library, https://www.jfklibrary.org/archives/other-resources/john-f-kennedy-speeches/pocatello-id-19600906; "Remarks of Senator John F. Kennedy, Milwaukee, Wisconsin," April 3,

1960, Kennedy Library, https://www.jfklibrary.org/archives/other-resources/john-f-kennedy-speeches/milwaukee-wi-19600403; "Remarks by Senator John F. Kennedy at San Francisco International Airport, San Francisco, California," September 3, 1960, Kennedy Library, https://www.jfklibrary.org/archives/other-resources/john-f-kennedy-speeches/san-francisco-19600903-airport.

9. Dwight D. Eisenhower, "Farewell Address," January 17, 1961, University of Virginia Miller Center, https://millercenter.org/the-presidency/presidential-speeches/january-17-1961-farewell-address.

10. John F. Kennedy, "Inaugural Address," January 20, 1961, University of Virginia Miller Center, https://millercenter.org/the-presidency/presidential-speeches/january-20-1961-inaugural-address; "Address in the Assembly Hall at the Paulskirche, Frankfurt," June 25, 1963, Kennedy Library, https://www.jfklibrary.org/asset-viewer/archives/JFKPOF/045/JFKPOF-045-023.

11. Waud, "Note"; Schlesinger, *Thousand Days*, 1005; "The Peculiar Recession," *Wall Street Journal*, November 25, 1960; "Nation Slips into 4th Postwar Recession; but Slide Is Less Severe Than Others," *Washington Post*, December 25, 1960; Dallek, *Unfinished Life*, 278–82.

12. Schlesinger, *Thousand Days*, 1006–7; Sorensen, *Kennedy*, 400–403; Arthur J. Goldberg, "The Challenge of 'Industrial Revolution II,'" *New York Times*, April 2, 1961.

13. "Goldberg Estimates Total Joblessness at 5.5 Million," *Los Angeles Times*, March 5, 1961; "Plan to Raise Job Aid Told by Goldberg," *Chicago Daily Tribune*, February 12, 1961; "Goldberg Clashes with Business Aides in Ohio," *New York Times*, February 12, 1961; "Jobless Tour Worthwhile, Kennedy Says," *Chicago Daily Tribune*, February 13, 1961.

14. Wilson, *Communities Left Behind*, 53–56; "President Signs Bill to Create Needy Area Jobs," *New York Times*, May 2, 1961; "Jobless Aid Bill Passed by Congress," *Washington Post*, March 23, 1961. The president had signed the supplementary unemployment compensation bill, and amendments to the farm bill, prior to signing Douglas's S. 1. The first bill he signed as president was a measure to commemorate Lincoln's first inaugural speech. "Kennedy Signs First Bill," *Christian Science Monitor*, March 2, 1961.

15. "Statement by the President upon Signing the Manpower Development and Training Act," March 15, 1962, American Presidency Project, https://www.presidency.ucsb.edu/documents/statement-the-president-upon-signing-the-manpower-development-and-training-act; US Bureau of Labor Statistics, *Technological Trends in Major American Industries*.

16. Executive Order 10918, "Establishing the President's Advisory Committee on Labor Management Policy," February 16, 1961, Center for Regulatory Effectiveness, http://thecre.org/fedlaw/legal12a/eo10918.htm; US Departments of Commerce and Labor, *Report of the President's Advisory Committee*, 1–11.

17. US Departments of Commerce and Labor, *Report of the President's Advisory Committee*, 1–11; "Labor-Business Panel Joblessness Report Is Non-specific, Aims at Gaining Support," *Wall Street Journal*, January 12, 1962.

18. "Ewan Clague, 90, U.S. Labor Official," *New York Times*, April 15, 1987; US Bureau of Labor Statistics, *Technological Trends in 36 Major American Industries*, 6–10, 16–17, 54–56.

19. US Bureau of Labor Statistics, *Technological Trends in 36 Major American Industries*, 6–10, 16–17, 54–56.

20. Senate Subcommittee on Employment and Manpower, *Nation's Manpower Revolution*, May 20, 1963, 1–3; June 6–7, 1963, 574–75.

21. Senate Subcommittee on Employment and Manpower, *Nation's Manpower Revolution*, May 20, 1963, 1–3; June 6–7, 574–75, 1751–69; "Walter Heller, 71, Economic Adviser in 60's, Dead," *New York Times*, June 17, 1981, https://www.nytimes.com/1987/06/17/obituaries/walter-heller-71-economic-adviser-in-60-s-dead.html?smid=nytcore-ios-share; Heller, New Dimensions, 63–64; "Challenge Interview"; Stricker, *American Unemployment*, 75–78.

22. Steven Greenhouse, "W. Willard Wirtz, Labor Chief, Dies at 98," *New York Times,* April 25, 2010; "Willard Wirtz, Labor Secretary for JFK and LBJ, Dies at 98," *Washington Post*, April 25, 2010; A. H. Raskin, "As Wirtz Sees His Basic Job," *New York Times*, November 11, 1962.

23. Senate Subcommittee on Employment and Manpower, *Nation's Manpower Revolution*, May 20, 1963, 5–11.

24. Schlesinger, *Thousand Days*, 632–33; Waterhouse, *Lobbying America*, 22–23.

25. Senate Subcommittee on Employment and Manpower, *Nation's Manpower Revolution*, May 20, 1963, 65, 84–85.

26. Senate Subcommittee on Employment and Manpower, *Nation's Manpower Revolution*, May 20, 1963, Wirtz testimony and responses, 4–71, and Hodges testimony, 72–98; US Department of Labor, "Summary Memorandum Describing the Legislative History, Purposes, and Composition of the National Commission on Technology, Automation, and Economic Progress," August 19, 1964, box 19, Howard Bowen Papers, University of Iowa Archives.

27. Senate Subcommittee on Employment and Manpower, *Nation's Manpower Revolution*, May 21, 1963, 113–50. Theobald separately laid out his more expansive vision of a guaranteed income in two separate volumes: *The Challenge of Abundance* and *The Rich and Poor*. He expressed his ideas in a series of articles that appeared in the May 11, 1963, issue of the *Nation*, entitled "An Economic Security Plan" (402–5), "Financing the Plan" (405–7), and "The Plan in Operation" (407–12).

28. Senate Subcommittee on Employment and Manpower, *Nation's Manpower Revolution*, May 22, 1963, 208–57, September 19–20, 1963, 1511–62.

29. Druckman, *Wayne Morse*, 11–32; "Presidential Commission on Automation," *Congressional Record*, 88th Cong., 1st sess., July 24, 1963, vol. 109, pt. 10, 13170–72.

30. "Remarks of Senator Jacob Javits," *Congressional Record*, 88th Cong., 1st sess., April 10, 1963, vol. 109, pt. 5, 6224–26.

31. "Jacob K. Javits Oral History Interview," April 26, 1966, Kennedy Library, https://archive1.jfklibrary.org/JFKOH/Javits,%20Jacob%20K/JFKOH-JKJ-01/JFKOH-JKJ-01-TR.pdf; Javits, *Javits*, 266.

32. Townsend, "Human Equation," 16; Waterhouse, *Lobbying America*, 22–23; Sorensen, *Kennedy*, 438–43; "Drafting Problems," 817; "Kennedy Expected to Give Rail Request to Congress Today," *Washington Post*, July 23, 1963; "Railroad Rules Dispute: Message from the President," *Congressional Record*, 88th Cong., 1st sess., July 22, 1963, vol. 109, pt. 10, 13004–7; "Kennedy Asks Congress for Rail Legislation," *Los Angeles Times*, July 23, 1963.

33. "Kennedy Expected"; "Railroad Rules Dispute," 13004–7; "Remarks of Senator Javits on the Railroad Rules Dispute," *Congressional Record*, 88th Cong., 1st sess., July 25, 1963, vol. 109, pt. 10, 13382; David Halberstam, "The Expensive Education of McGeorge Bundy," *Harper's*, July 1969. David Halberstam actually coined the phrase "the best and the brightest" to describe McGeorge Bundy. But he noted that the same term could be used to "describe the entire group [of Kennedy advisers] as it swept so confidently into Washington." Halberstam, *Best and the Brightest*, xix; White, *In Search of History*, 492–93.

34. "Presidential Commission on Automation," *Congressional Record*, 88th Cong., 1st sess., July 24, 1963, vol. 109, pt. 10, 13170–71.

35. In addition to Javits and Morse, the other cosponsors were Senators Edward V. Long (D-MO) and Hiram Fong (R-HW). "Presidential Commission on Automation," 13170–71, 13389; "Remarks of Senator Javits," 13107; Javits, *Javits*, 392. By 1931 the *Congressional Almanac*, intrigued that President Hoover had appointed some 30 commissions during his administration, tabulated that 492 commissions had been appointed since Theodore Roosevelt first became intrigued with the British Royal Commission model (see George B. Galloway, "Presidential Commissions," *Congressional Quarterly*, May 28, 1931). In his exhaustive study of the types of presidential commissions from Truman through Nixon's first term, Thomas R. Wolanin documents 40 advisory commissions preceding the appointment of Johnson's on automation, and there would be another 25 during Johnson's tenure. See Wolanin, *Presidential Advisory Commissions*, 205–15.

36. "Representative James G. Fulton Dead, Dean of Pennsylvania Delegation," *New York Times*, October 7, 1971. This obituary notes that "although his 27th congressional district in Pittsburgh had about 35,000 more Democrats than Republicans, Mr. Fulton was consistently re-elected." "James G. Fulton: Congressman with a Personal Touch," *Pittsburgh Post-Gazette*, February 18, 2015, https://newsinteractive.post-gazette.com/thedigs/2015/02/18/james-g-fulton-congressman-with-a-personal-touch/; "Federal Commission on Automation," *Congressional Record*, 88th Cong., 1st sess., September 12, 1963, vol. 109, pt. 12, 16923–25.

37. "Kennedy Signs Bill Averting a Rail Strike," *New York Times*, August 29, 1963; Barber, *Marching on Washington*, 108–40, 141–78; Forstater, "Civil Rights to Economic Security."

38. NBCUniversal Archives, "August 28, 1963: March on Washington," YouTube video, 28:34 (Reuther's speech begins at 6:31), August 6, 2013, https://www.youtube.com/watch?v=MXa_dopI3VQ; "200,000 Attend Peaceful D.C. March, Kennedy Says Nation Can Be Proud, President Signs Bill Averting Rail Strike," *Washington Post*, August 29, 1963; Lichtenstein, *Walter Reuther*, 370–88.

39. "Kennedy Signs Bill"; Civil Rights Act of 1964, PL 88-352, 88th Cong., 2nd sess., July 2, 1964.

40. "Extension of Remarks," *Congressional Record*, 88th Cong., 1st sess., November 29, 1963, vol. 109, pt. 17, 22963–64.

41. Richard L. Madden, "Fighter for Civil Rights," *New York Times*, December 27, 1976; "Commission on the Application of Technology to Community and Manpower Needs," *Congressional Record*, 88th Cong., 1st sess., November 8, 1963, vol. 109, pt. 16, 21452–56.

42. George Meany, "Remarks at AFL-CIO Convention, New York," November 14, 1963, Kennedy Library, https://www.jfklibrary.org/asset-viewer/archives/JFKPOF/048/JFKPOF-048-008; Damon Stetson, "Rockefeller Opposes Right to Work Laws," *New York Times*, November 15, 1963.

43. Lyndon B. Johnson, conversation with Walter Reuther, November 23, 1963, University of Virginia Miller Center, https://millercenter.org/the-presidency/secret-white-house-tapes/conversation-walter-reuther-november-23-1963; Cormier and Eaton, *Reuther*, 371, 374–75, 380–81.

44. Caro, *Passage of Power*, 545–51; Lyndon B. Johnson, "Annual Message to the Congress on the State of the Union," January 8, 1964, American Presidency Project, https://www.presidency.ucsb.edu/documents/annual-message-the-congress-the-state-the-union-25.

7. The Automation Commission

1. The National Medal of Science was created by an act of Congress on August 25, 1959, during the Eisenhower administration. President Kennedy was the first to confer the award. See John F. Kennedy, "Executive Order 10961—Providing Procedures for the Award of the National Medal of Science," August 21, 1961, American Presidency Project, https://www.presidency.ucsb.edu/documents/executive-order-10961-providing-procedures-for-the-award-the-national-medal-science; Lyndon B. Johnson, "Remarks at the National Medal of Science Award Ceremony," January 13, 1964, American Presidency Project, http://www.presidency.ucsb.edu/documents/remarks-the-national-medal-science-award-ceremony; Zachary, *Endless Frontier*, 393–97; "Johnson Presents Science Medals," *Washington Post*, January 14, 1964.

2. Johnson, "Remarks"; Conway and Siegelman, *Dark Hero*, 322–28; "Dr. Wiener, World Famed Scientists, Dies," *Los Angeles Times*, March 19, 1964; "High Honor Won by 3 at M.I.T.," *Boston Globe*, January 14, 1964.

3. Carl Solberg, *Hubert Humphrey*, 219; White, *Making of the President 1960*, 29–34; Humphrey, *Education of a Public Man*, 335; "Commission on Automation, Technology, and Employment," *Congressional Record*, 88th Cong., 2nd sess., January 15, 1964, vol. 110, pt. 1, 498–501.

4. "Commission on Automation," 498–99; Heady, "Reports of the Hoover Commission."

5. Memorandum from Phillip S. Hughes, Assistant Director for Legislative Affairs, to Myer Feldman, Counsel to the President, March 5, 1964, box 392, LBJ Presidential Library (hereafter LBJ Library). For reference to Phillip S. (Sam) Hughes, see "The President's News Conference of February 26, 1966," in *Public Papers*, 216–17.

6. Lyndon Johnson to the President of the Senate and the Speaker of the House of Representatives, March 9, 1964, box 824, John Macy Papers, LBJ Library; "To Establish a National Commission on Automation and Technological Progress, S. 2623," *Congressional Record*, 88th Cong., 2nd sess., March 10, 1964, vol. 110, pt. 2, 4783–86, 4985.

7. "To Establish a National Commission," 4783–86.

8. Members of the Ad Hoc Committee on the Triple Revolution are listed in their cover letter to Johnson, March 22, 1964; see Linus Pauling and the International

Peace Movement Published Papers and Official Documents, Special Collections and Archives Research Center, Oregon State University (hereafter Pauling Papers), http://scarc.library.oregonstate.edu/coll/pauling/peace/papers/1964p.7-01. html, p. 3; Harrington, *Other America*, 13, 27, 79; "Michael Harrington: Warrior on Poverty," *New York Times*, June 19, 2009.

9. For the text of the "The Triple Revolution," see "The Triple Revolution," March 22, 1964, Pauling Papers, http://scarc.library.oregonstate.edu/coll/pauling/peace/papers/1964p.7-01.html; "New Pay Idea Urged for Era of Computers," Chicago Tribune, March 23, 1964.

10. See "Triple Revolution"; "Guaranteed Income Asked for All, Employed or Not," *New York Times*, March 23, 1964; Lee C. White to W. H. Ferry, April 6, 1964, Pauling Papers, http://scarc.library.oregonstate.edu/coll/pauling/peace/papers/1964p.7-04.html; "Cybernation Revolution," *Washington Post*, March 28, 1964; "Ad Hokum," *Baltimore Sun*, March 26, 1964; Frank C. Porter, "Wirtz Hits Guaranteed Income Plan," *Washington Post*, March 24, 1964; House Select Subcommittee on Labor, *Hearings on H.R. 10310*, 125–35.

11. Michael D. Reagan, "For a Guaranteed Income," *New York Times Magazine*, June 7, 1964.

12. "The Honorable Elmer J. Holland of Pennsylvania," *Congressional Record*, 90th Cong., 2nd sess., July 30, 1968, vol. 114, pt. 18, 24273–75. "Elmer Joseph Holland (1894–1968)," Biographical Directory of the United States Congress, accessed August 10, 2021, https://bioguide.congress.gov/search/bio/H000717.

13. House Select Subcommittee on Labor, *Hearings on H.R. 10310*, 104–6 (Reuther), 116–19 (Watson), 110–12 (Beirne).

14. "Clarence J. Brown," Biographical Directory of the United States Congress, accessed August 10, 2021, http://bioguide.congress.gov/scripts/biodisplay. pl?index=B000909; House Select Subcommittee on Labor, *Hearings on H.R. 10310*, 15–20.

15. House Select Subcommittee on Labor, *Hearings on H.R. 10310*, 18–26, 79–85; US Bureau of Labor Statistics, *Technological Trends in 36 Major American Industries*.

16. Senate Labor and Public Welfare Subcommittee on Employment and Manpower, *Nation's Manpower Revolution*, 3388–98, 3400–3412 (Clark quote appears on 3405).

17. "Congressman Holland Speaking on H.R. 11611," *Congressional Record*, 88th Cong., 2nd sess., July 21, 1964, vol. 110, pt. 12, 16462–68; "National Commission on Technology, Automation, and Economic Progress," *Congressional Record*, 88th Cong., 2nd sess., July 31, 1964, vol. 110, pt. 13, 17573–79, and August 5, 1964, vol. 110, pt. 14, 18177, 18194.

18. List of invitees to bill-signing ceremony, n.d.; telegrams received as RSVPs, August 18, 1964; and text of telegram of invitation, August 18, 1964, all in box 392, LBJ Library; "Remarks of the President at the Signing of H.R. 116111 Commission on Automation," August 19, 1964, box 10, LBJ Library; White, *Making of the President 1964*, 237–45. The quote from Eliot Janeway appears in White, *Making of the President*, 244.

19. List of invitees and attendees at bill-signing ceremony, telegrams received as RSVPs, and text of telegram of invitation, box 392, LBJ Library; "Remarks of the President"; Lyndon B. Johnson, "Remarks upon Signing Bill Creating the National

Commission on Technology, Automation, and Economic Progress," August 19, 1964, American Presidency Project, https://www.presidency.ucsb.edu/documents/remarks-upon-signing-bill-creating-the-national-commission-technology-automation-and.

20. Waterhouse, Lobbying America, 25.

21. "Reuther Sees Threat in Push Button Option," *Baltimore Sun*, December 5, 1954.

22. Johnson, "Remarks upon Signing Bill"; Dallek, Lyndon B. Johnson, 155. GDP in 1964 would grow at 5.8 percent; see Federal Reserve Economic Data, "Gross Domestic Product," Federal Reserve Bank of St. Louis, accessed October 27, 2021, https://fred.stlouisfed.org/series/GDP; Lyndon B. Johnson, State of the Union address, January 8, 1964, University of Virginia Miller Center, https://millercenter.org/the-presidency/presidential-speeches/january-8-1964-state-union.

23. Charles E. Walker to Walter Jenkins, September 28, 1964, box 393, LBJ Library; Bart Barnes, "Charles E. Walker, Tax Lobbyist for GOP and Big Business, Dies at 91," *New York Times*, June 29, 2015; Jenkins, note to Walker, September 29, 1964, box 393, LBJ Library; Waterhouse, *Lobbying America*, 229–32.

24. See letters per nominees in box 824, LBJ Library; Senator Philip A. Hart, memorandum to Johnson, July 31, 1964, box 824, LBJ Library; Congressman Elmer Holland to Wilson McCarthy at the Peace Corps, June 3, 1963, recommending that Robert Ryan be appointed to the commission, box 393, LBJ Library; Luther B. Hodges, Secretary of Commerce, memorandum to Ralph Dungan, Special Assistant to the President, August 14, 1964, box 824, LBJ Library. According to Hodges's memo, "There has been no consensus" in the Department of Commerce as to nominees for the commission. Letters recommending candidates and lists of candidates compiled by Dungan, John Macy, and John Clinton, including Macy's summary list, and Walter W. Heller's August 14, 1964, summary memo to Ralph Dungan, are in box 824, LBJ Library. The reaction of Robert Ryan is taken from an article of unidentified origin in box 393, LBJ Library.

25. An undated "Note for the Files" was sent by John Clinton to Stan Ruttenberg, box 393, LBJ Library.

26. "Bowen Loses Rank of Dean at Illinois University," *Baltimore Sun*, December 29, 1950; "The Bowen Case Is Closed," *Chicago Tribune*, December 31, 1950.

27. "Dungan, Ralph Anthony, Jr., 1923–2013," DiscoverLBJ, accessed October 27, 2021, https://www.discoverlbj.org/item/dunganr; telegrams to individual nominees to serve on the commission, November 14, 1964, and a draft press release announcing the nominees, November 14, 1964, box 393, LBJ Library; "President Names 14 to Automation Unit," *New York Times*, November 15, 1964; Ralph A. Dungan, memorandum to John Clifton, October 22, 1964, and LBJ's cover slip to Walter Jenkins, September 11, 1964, box 824, LBJ Library; President's Daily Diary entry, January 6, 1966, President's Daily Diary Collection, LBJ Library, https://www.discoverlbj.org/item/pdd-19650106 (see p. 7); "Automation Panel Approved," *New York Times*, January 26, 1965.

28. "Senators Affirm Okun as Adviser to Johnson," *New York Times*, January 28, 1965; "Automation Unit Begins Work," *New York Times*, January 30, 1965; "Anna Rosenberg Hoffman Dead: Consultant and 50's Defense Aid," *New York Times*, May 10, 1983.

29. Requirements for selection to the commission are specified in section 3 of the Act to Establish a National Commission on Technology, Automation, and Economic Progress, PL 88-444, 88th Cong., 2nd sess., August 19, 1964; "Kennedy Urged: Act Now on Automation," *Chicago Tribune*, September 3, 1963; Joseph A. Beirne to Senator Lister Hill, March 16, 1964, box 393, LBJ Library; Beirne to Congressman Adam Clayton Powell, March 16, 1964, box 393, LBJ Library; Senate Subcommittee on Employment and Manpower, *Nation's Manpower Revolution*, October 28, 1963, 3202–5.

30. "Wagner Asks WPA to Aid Jobless," *New York Amsterdam News*, December 14, 1963; Shales, *Great Society*, 151; "NUL Foresees Experience in Disillusionment," *Chicago Defender*, September 8, 1962; Dallek, *Flawed Giant*, 415.

31. Bunanos, *Instant*, 13; Phipps, "Early History"; House Select Subcommittee on Labor, *Hearings on H.R. 10310*, 117–19; "John I. Snyder Jr.," *New York Times*, April 26, 1965; "New Factory of U.S. Industries to Make Automation Equipment," *New York Times*, October 21, 1961; *Nuclear Power Economics*, 16–22.

32. Waters, *Daniel Bell*, 106–7; Solow, "Technical Change." In 1961, Solow was given the John Bates Clark Medal by the American Economics Association. The medal is awarded annually to an American economist under the age of forty who is judged to have made the most significant contribution to economic thought and knowledge. See "John Bates Clark Medal," American Economics Association, accessed October 27, 2021, https://www.aeaweb.org/about-aea/honors-awards/bates-clark.

33. "Kennedy Names Three to Settle Rail Dispute," *Boston Globe*, September 6, 1963; "Benjamin Aaron, an Expert in Labor Law, Dies at 91," *New York Times*, August 31, 2007.

34. Ralph Dungan, note to John B. Clinton, October 22, 1964; Gardner Ackley, note to Clinton, forwarding a draft of nomination to the commission to be sent by the president, October 26, 1964; and Clinton to Howard Bowen, December 8, 1964, all in box 824, LBJ Library; John W. Macy Jr., memorandum to Johnson, February 1, 1965; and White House press release on president's intention to announce Garth Mangum to serve as executive secretary to the commission, n.d. (his nomination was sent to the Senate on February 8, 1965), both in box 305, LBJ Library; President's Daily Diary entry, February 2, 1965, President's Daily Diary Collection, LBJ Library, https://www.discoverlbj.org/item/pdd-19650202; "Obituary: Garth Mangum," *Salt Lake City Tribune*, June 8, 2014; Senate Committee on Labor and Public Works, *Dr. Garth Mangum*; Garth Mangum to Mary Rawling, Office of Jack Valenti, January 22, 1965, box 392, LBJ Library.

35. Bowen apparently offered the position of executive secretary to Neil W. Chamberlain, an economics professor emeritus at Yale. Like Bowen, Chamberlain had written on the broader social responsibilities of labor and business; see Neil W. Chamberlain to Howard R. Bowen, November 24, 1964, box 19, Howard Bowen Papers, University of Iowa Archives (hereafter Bowen Papers); Bowen to Garth Mangum, January 18, 1965; Bowen to Mangum, January 11, 1965; Mangum to Bowen, January 14, 1965, all in box 19, Bowen Papers; Frank Lynn, interview by the author, Grand Haven, MI, August 10, 2017.

36. Howard R. Bowen to Garth Mangum, January 18, 1965; Bowen to Mangum, January 11, 1965; Mangum to Bowen, January 14, 1965, all in box 19, Bowen Papers.

37. Garth L. Mangum, memorandum to commission members, June 8, 1965, box 19, Bowen Papers. At the February 18 meeting of the commission, witnesses included J. Herbert Hollomon, the assistant secretary for science and technology, and Leon Greenberg, the assistant commissioner for productivity and technological development with the Bureau of Labor Statistics. Howard R. Bowen to J. Herbert Hollomon, February 24, 1965; Leon Greenberg to Bowen, February 24, 1965, all in box 19, Bowen Papers; Bowen to Lyndon Johnson, July 1, 1965, box 392, LBJ Library.

38. Jack Valenti, note to Lyndon Johnson, December 10, 1964, box 392, LBJ Library; Johnson to Harold R. Bowen, February 13, 1965, box 19, Bowen Papers.

39. Act to Establish a National Commission.

8. Bold Solutions

1. Lyndon B. Johnson, "Remarks in Athens at Ohio University," May 7, 1964, American Presidency Project, https://www.presidency.ucsb.edu/documents/remarks-athens-ohio-university; Lyndon B. Johnson, "Remarks at the University of Michigan," May 22,1964, American Presidency Project, https://www.presidency.ucsb.edu/documents/remarks-the-university-michigan; Lyndon B. Johnson, "Annual Message to the Congress on the State of the Union," January 4, 1965, American Presidency Project, https://www.presidency.ucsb.edu/documents/annual-message-the-congress-the-state-the-union-26; Lyndon B. Johnson, inaugural address, January 20, 1965, American Presidency Project, https://www.presidency.ucsb.edu/documents/the-presidents-inaugural-address.

2. Shales, *Great Society*, 1–16; National Commission on Technology, Automation, and Economic Progress, *Statements*, xiii.

3. Terborgh, *Automation Hysteria*, 3–11; "U.S. Agencies Biggest Uses of Computers," *Los Angeles Times*, December 9, 1965; "Automated Factories: Computers Take Over Bigger Number of Jobs on the Production Line," *Wall Street Journal*, April 10, 1964.

4. National Commission on Technology, Automation, and Economic Progress, *Statements*, 45–53.

5. Howard R. Bowen to Lyndon Johnson, July 1, 1965, box 392, LBJ Presidential Library (hereafter LBJ Library); Rodgers, *Atlantic Crossings*, 93–94; Bell, "Government by Commission"; Garth Mangum, "Commission's Letter Requesting Statements," n.d., National Commission on Technology, Automation, and Economic Progress, *Statements*, 1.

6. National Commission on Technology, Automation, and Economic Progress, *Statements*, xiii, 242–43.

7. National Commission on Technology, Automation, and Economic Progress, *Statements*, 13–22, 123, 271.

8. National Commission on Technology, Automation, and Economic Progress, *Statements*, 87–88, 122, 171.

9. National Commission on Technology, Automation, and Economic Progress, *Statements*, 27, 91–112, 121, 123, 171.

10. National Commission on Technology, Automation, and Economic Progress, *Statements*, 185–87, 227–32, 275, 279–80, 283.

11. "Ray R. Eppert," *New York Times*, July 22, 1986; Jack Valenti to Ray R. Eppert, September 27, 1965, box 393, LBJ Library; National Commission on Technology, Automation, and Economic Progress, *Statements*, 63–65.

12. National Commission on Technology, Automation, and Economic Progress, *Statements*, 305–7.

13. National Commission on Technology, Automation, and Economic Progress, *Statements*, 149–56, 157–60, 233–38, 285–292, 293–302.

14. National Commission on Technology, Automation, and Economic Progress, *Statements*, 3–6, 7–12.

15. National Commission on Technology, Automation, and Economic Progress, *Statements*, 127–35, 253–60.

16. Bell, *Post-industrial Society*, 209.

17. Frank Lynn, interview by the author, Grand Haven, MI, August 10, 2017.

18. Bell, *Post-industrial Society*, 209; National Commission on Technology, Automation, and Economic Progress, *Employment Impact*, 31–44.

19. Bell, *Post-industrial Society*, 209.

20. Bell, *Work and Its Discontents*, 46–54; Diamond, "Edwin Mansfield's Contributions"; Mansfield, "Size of the Firm," 573; Mansfield, "Technical Change"; National Commission on Technology, Automation, and Economic Progress, *Employment Impact*, 2:128–30.

21. Solow, "Contribution to the Theory"; Solow, "Technical Change."

22. National Commission on Technology, Automation, and Economic Progress, *Employment Impact*, 43; Act to Establish a National Commission on Technology, Automation, and Economic Progress, PL 88-444, 88th Cong., 2nd sess., August 19, 1964, 462.

23. Malone, *Intel Trinity*, 13–18; Denning, "Science of Computing," 530.

24. In 1979, manufacturing employment reached its postwar peak of 19,301,000 employees, never to be topped since; see "Employment, Hours, and Earnings from the Current Employment Statistics Survey (National)," Bureau of Labor Statistics, accessed January 12, 2021, https://data.bls.gov/pdq/SurveyOutputServlet. Willard Wirtz, Secretary of Labor, memorandum to Lyndon B. Johnson, June 2, 1965, box 6, LBJ Library; see also "Databases, Tables & Calculators by Subject: Unemployment Rate," Bureau of Labor Statistics, accessed January 12, 2021, https://data.bls.gov/pdq/SurveyOutputServlet; Lyndon B. Johnson, "Remarks upon Signing Bill Creating the National Commission on Technology, Automation, and Economic Progress," August 19, 1964, American Presidency Project, https://www.presidency.ucsb.edu/documents/remarks-upon-signing-bill-creating-the-national-commission-technology-automation-and; Wilbur J. Cohen, Secretary of Health, Education, and Welfare, memorandum to Johnson, August 11, 1965, box 6, LBJ Library.

25. Cohen, memorandum to Johnson, August 11, 1965.

26. Howard R. Bowen to John W. Macy Jr., September 27, 1965, box 192, LBJ Library; Macy, memorandum to Gardner Ackley, Charles L. Schultze, Willard Wirtz, John T. Connor, and Bill Moyers, October 20, 1965, box 192, LBJ Library. Macy circulated Bowen's request to CEA director Gardner Ackley, Labor Secretary Wirtz, Schultze at the Bureau of the Budget Director, Commerce Secretary John Connor, and Press Secretary Bill Moyers. Macy to Bowen, October 20, 1965, box 192, LBJ Library.

27. "Government Steps into Automation," *Boston Globe*, November 21, 1965; "Expect Jobs Gain despite Machines," *Chicago Tribune*, November 22, 1965; "Federal Panel Discounts Job Peril in Automation," *New York Times*, December 23, 1965; "Perspective on Automation," *New York Times*, December 25, 1965; Bell, "Government by Commission."

28. "Consensus Sought on Automation," *New York Times*, January 9, 1966; "Commission Splits over Automation," *Baltimore Sun*, December 24, 1965; "Panel Split on Impact of Automatic Growth," *Washington Post*, December 24, 1965.

29. National Commission on Technology, Automation, and Economic Progress, *Technology*, 1–5.

30. National Commission on Technology, Automation, and Economic Progress, *Technology*, 21–31, 110–11.

31. National Commission on Technology, Automation, and Economic Progress, *Technology*, 48–56, 110–11.

32. National Commission on Technology, Automation, and Economic Progress, *Technology*, 17–26, 46.

33. National Commission on Technology, Automation, and Economic Progress, *Technology*, 23–31, 110; "Education: More School?," *New York Times*, March 1, 1964; "Urge 2 Years More of Free Education," *Chicago Tribune*, January 2, 1964.

34. National Commission on Technology, Automation, and Economic Progress, *Technology*, 38–41, 110.

35. National Commission on Technology, Automation, and Economic Progress, *Technology*, 73–94; "Suburban Migration," *Congressional Quarterly Researcher*, July 20, 1960, http://library.cqpress.com/cqresearcher/document.php?id=cqresrre1960072000.

36. National Commission on Technology, Automation, and Economic Progress, *Technology*, 105–6.

37. National Commission on Technology, Automation, and Economic Progress, *Technology*, 106–8, 112.

38. "GNP Passed $675 Billion in 1965 for 7.5% Boost," *Washington Post*, January 6, 1966; "N.I.C.B. Sees 727 Billion GNP in 1966," *Chicago Tribune*, January 3, 1966; "NAM Predicts $715-Billion GNP," *Washington Post*, January 9, 1966.

39. Edward L. Sherman, memorandum to John W. Macy Jr., January 19, 1966, box 824, LBJ Library; Sherman, memorandum to Bill Moyers (with handwritten return note from Moyers), John Macy file, box 305, LBJ Library; Hayes Redmon, note to Moyers, January 24, 1966, box 392, LBJ Library; Jack Valenti, note to Lyndon Johnson, January 27, 1966, 7:05 (no a.m./p.m. designation), box 392, LBJ Library.

40. "Advance Press Release to P.M. Newspapers," February 2, 1966, box 192, Papers of Lyndon Baines Johnson, LBJ Library; "16 Inches of Snow Stalls Area," *Washington Post*, January 31, 1966; Bell, "Government by Commission," 5; Howard R. Bowen to Lyndon Johnson, February 2, 1966; Lawrence E. Levinson, Special Counsel, note to Joseph Califano, Special Assistant to the President, February 14, 1966; Arthur M. Okun, Economic Adviser to the President, memorandum to Larry Levinson, February 12, 1966; Califano, note to Johnson, February 21, 1966, all in box 392, LBJ Library.

41. Senator Jacob Javits, "Remarks," *Congressional Record*, 89th Cong., 2nd sess., February 3, 1966, vol. 112, pt. 2, 2041; "Remarks of Senator Joseph Clark,"

Congressional Record, 89th Cong., 2nd sess., February 9, 1966, vol. 112, pt. 2, 2732–33; "Remarks of Congressman Joe Waggonner," *Congressional Record*, 89th Cong., 2nd sess., February 8, 1966, vol. 112, pt. 2, 2501.

42. "Minimum Family Income Is Urged," *New York Times*, February 4, 1966; "Negative Income Tax Idea Urged as Aid to Needy," *New York Times*, February 6, 1966; Roscoe Drummond, "Labor, Business See Eye to Eye," *Boston Globe*, February 12, 1966; "Task Force Proposes Job and Income Props," *Washington Post*, February 4, 1966; "Government Panel Issues Report Skirting Question of Automation's Impact on Jobs," *Wall Street Journal*, February 4, 1966.

43. Raskin, "Great Society," 556–58; A. H. Raskin, "Pattern for Tomorrow's Industry," *New York Times*, December 18, 1955; "A. H. Raskin, 82, Times Reporter and Editor, Dies," *New York Times*, December 23, 1993.

Epilogue

1. Ford, *Rise of the Robots*, 1–12.

2. Ratner, "Especially about the Future," 99; Lichtenstein, *Contest of Ideas*, 129–33.

3. These data show the unemployment rate to be well beneath 4 percent in April and June of 1929, then skyrocketing to 25 percent by April 1932; see Federal Reserve Economic Data, "FRED Graph," Federal Reserve Bank of St. Louis, accessed October 27, 2021, https://fred.stlouisfed.org/graph/?g=jS03. As early as 1971, Robert Solow joined Edward Denison and George Perry in observing that although in the period 1956–70, manufacturing employment continued to go up, productivity rates were declining. Also see the introduction, note 22.

4. Perlstein, *Nixonland*, 163–66; Zelizer, *Fierce Urgency of Now*, 228–61.

5. Martin Luther King Jr., "The Three Evils of Society," August 31, 1967, https://www.scribd.com/doc/134362247/Martin-Luther-King-Jr-The-Three-Evils-of-Society-1967.

6. "Guardsman Kill 4 Kent State Students," *Baltimore Sun*, May 5, 1970; "Kent State Is a Scene of Unreality," *Chicago Tribune*, May 5, 1970; "Four Kent State Students Killed by Troops," *New York Times*, May 5, 1970; "Troops Kill Four Students in Anti-war Riot at Ohio College," *Los Angeles Times*, May 5, 1970; "Man Dies as Bomb Rips Math Center," *New York Times*, August 25, 1970; O'Neill, *Coming Apart*; "Walter Reuther: Union Pioneer with Broad Influence Far beyond the Field of Labor," *New York Times*, May 11, 1970.

7. "Changing Productivity Trends," FRSF Economic Letter, August 31, 2007, Federal Reserve Bank of San Francisco, https://www.frbsf.org/economic-research/publications/economic-letter/2007/august/productivity-trends/; US Congressional Budget Office, *Productivity Slowdown*, 6–9; Federal Reserve Economic Data, "All Employees, Manufacturing," Federal Reserve Bank of St. Louis, accessed March 21, 2021, https://fred.stlouisfed.org/series/MANEMP. At the time the Worker Adjustment and Retraining Act passed in 1988, the manufacturing employment rate hovered at 18 million; by 2020 it was approximately 12 million. National Commission on Technology, Automation, and Economic Progress, *Technology*, 89; Cowie, *Stayin' Alive*, 47–48; Clayton Fritchey, "The Big Grind of the Assembly Line," *Chicago Tribune*, April 9, 1972.

8. Representative Augustus Hawkins, "Remarks," *Congressional Record*, 93rd Cong., 2nd sess., June 26, 1974, vol. 120, pt. 16, 21278–79.

9. Senator Hubert Humphrey, "Remarks," *Congressional Record*, 93rd Cong., 2nd sess., August 22, 1974, vol. 120, pt. 22, 29785–86.

10. House Committee on Banking, Housing, and Urban Affairs, *Full Employment and Balanced Growth Act*, 1–2, 130–42.

11. Cowie, *Stayin' Alive*, 265–78; Stein, *Pivotal Decade*, loc. 3593–763 of 8276, Kindle; Andelic, "Old Economic Rules"; Waterhouse, *Lobbying America*, 130–32; "Humphrey-Hawkins Full Employment Bill"; "Full Employment Bill Passes the Senate," *Chicago Tribune*, October 14, 1978.

12. "President Signs Symbolic Humphrey-Hawkins Bill," *Los Angeles Times*, October 28, 1978; "More Blacks Reported Jobless," *Boston Globe*, October 30, 1978; "In Minnesota, Liberal Ways Face Challenge," *Boston Globe*, October 28, 1978.

13. "Industrial Robots: Auto Workers Back Growing Use, Also Benefit," *Washington Post*, December 25, 1978; "How Mechanical Workers are Clanking in Industry," *Christian Science Monitor*, December 27, 1978; "More Blacks Reported Jobless," *Boston Globe*, October 30, 1978.

14. Trubek, *Voices*, 1.

15. Eugene Kim, "Jeff Bezos to Employees: 'One Day, Amazon Will Fail' but Our Job Is to Delay It as Long as Possible," CNBC, last updated November 27, 2018, https://www.cnbc.com/2018/11/15/bezos-tells-employees-one-day-amazon-will-fail-and-to-stay-hungry.html; Brynjolfsson and McAfee, *Second Machine Age*, loc. 36–180 of 5394, Kindle.

16. "Tribute to the Late President John Kennedy, Senator Saltonstall," *Congressional Record*, 88th Cong., 2nd sess., January 15, 1964, vol. 110, pt. 1, 480.

BIBLIOGRAPHY

Archives

Electronic archives of Presidents Herbert Hoover, Franklin D. Roosevelt, Harry S. Truman, Dwight D. Eisenhower, John F. Kennedy, and Lyndon B. Johnson.

Howard Bowen Papers, University of Iowa Archives, Iowa City, IA.

Papers of Lyndon Baines Johnson, LBJ Presidential Library, Austin, TX.

Printed Publications

Abramowitz, Alan I. *The Great Alignment: Race, Party Transformation, and the Rise of Donald Trump*. New Haven, CT: Yale University Press, 2018.

Acemoglu, Daron, and Restrepo, Pascual. "Automation and New Tasks: How Technology Displaces and Reinstates Labor." *Journal of Economic Perspectives* 33, no. 2 (Spring 2019): 3–30.

Allen, Frederick Lewis. *Only Yesterday*. New York: Open Road, 1931.

Allen, Leroy. "Technocracy—a Popular Summary." *Social Science* 8, no. 2 (April 8, 1933): 175–88.

Alter, Jonathan. *The Defining Moment: FDR's Hundred Days and the Triumph of Hope*. New York: Simon and Schuster, 2006.

Amsterdam, Daniel. "Toward a Civic Welfare State." In *Capital Gains: Business and Politics in Twentieth-Century America*, edited by Richard R. John and Kim Phillips-Fein, 43–58. Philadelphia: University of Pennsylvania Press, 2017.

Andelic, Patrick. "The Old Economic Rules No Longer Apply." *Journal of Policy History* 31, no. 1 (July 2019): 72–110.

"Area Redevelopment Bill Vetoed." In *CQ Almanac 1958*, 147–51. Washington, DC: Congressional Quarterly, 1959.

Bachmura, Frank T. "The Manpower Development and Training Act of 1962." *Journal of Farm Economics* 75, no. 1 (February 1963): 61–72.

Badger, Anthony J. *The First One Hundred Days*. New York: Hill and Wang, 2008.

Bailey, Martin Neil. "The Productivity Growth Showdown and Capital Accumulation." *American Economic Review* 71, no. 2 (May 1981): 326–31.

Bailey, Stephen Kemp. *Congress Makes a Law*. New York: Vintage Books, 1950.

Barber, Lucy G. *Marching on Washington: The Forging of an American Political Tradition*. Berkeley: University of California Press, 2002.

Barnard, Harry. *Independent Man: The Life of Senator James Couzens*. Detroit: Wayne State University Press, 1958. Kindle.

Baruch, Bernard M., and John M. Hancock. *Report on War and Post-war Adjustment Policies*. Washington, DC: US Government Printing Office, 1944.

Baxter, James Phinney. *Scientists against Time*. Cambridge: MIT Press, 1968.

Beardsley, William W., ed. *1945 Collier's Yearbook*. New York: P. F. Collier and Son, 1945.

Bell, Daniel. *The Coming of Post-industrial Society: A Venture in Social Forecasting*. New York: Basic Books, 1973.

Bell, Daniel. "Government by Commission." *Public Interest*, no. 43 (Spring 1966): 3–9.

Bell, Daniel. *Work and Its Discontents: The Cult of Efficiency in America*. Boston: Beacon, 1956.

Bernstein, Irving. *The Lean Years: A History of the American Worker*. Boston: Houghton Mifflin, 1960.

Beveridge, William H. *Full Employment in a Free Society: A Report*. London: George Allen and Unwin, 1944.

Biles, Roger. *Crusading Liberal: Paul H. Douglas of Illinois*. DeKalb: Northern Illinois University Press, 2002.

Bird, Kai, and Martin J. Sherwin. *American Prometheus: The Triumph and Tragedy of J. Robert Oppenheimer*. New York: Vintage Books, 2006. Kindle.

Bix, Amy Sue. *Inventing Ourselves Out of Jobs? America's Debate over Technological Unemployment, 1929–1981*. Baltimore: Johns Hopkins University Press, 2000.

Bix, Amy Sue. "Inventing Ourselves Out of Jobs? America's Depression Era Debate over Technological Unemployment." PhD diss., Johns Hopkins University, 1994.

Boyle, Kevin. *The UAW and the Heyday of American Liberalism*. Ithaca, NY: Cornell University Press, 1995.

Brewer, Mark D. "The Evolution and Alteration of American Party Coalitions." In *The Oxford Handbook of American Political Parties and Interest Groups*, edited by L. Sandy Maisel and Jeffrey M. Berry, 121–42. New York: Oxford University Press, 2010.

Brinkley, Alan. *The End of Reform: New Deal Liberalism in Recession and War*. New York: Vintage Books, 1995.

Bronk, Detley W. "National Science Foundation: Origins, Hopes, and Aspirations." *Proceedings of the National Academy of Sciences of the United States of America* 72, no. 8 (August 1975): 2839–42.

Brynjolfsson, Erik, and Andrew McAfee. *The Second Machine Age: Work, Progress, and Prosperity in a Time of Brilliant Technologies*. New York: W. W. Norton, 2014. Kindle.

Bunanos, Christopher. *Instant: A Cultural History of Polaroid*. New York: Princeton Architectural Press, 2012.

Bush, Vannevar. "Science: The Endless Frontier." *Transactions of the Kansas Academy of Science* 48, no. 3 (1945): 231–64.

Cain, Louis, and George Neumann. "Planning for Peace: The Surplus Prosperity Act of 1944." *Journal of Economic History* 44, no. 1 (1981): 129–35.

Caro, Robert A. *Master of the Senate: The Years of Lyndon Johnson*. New York: Vintage Books, 2002.

Caro, Robert A. *Means of Ascent: The Years of Lyndon Johnson*. New York: Vintage Books, 1990.

Caro, Robert A. *The Passage of Power: The Years of Lyndon Johnson*. New York: Alfred A. Knopf, 2012.

Caro, Robert A. *The Path to Power: The Years of Lyndon Johnson*. New York: Vintage Books, 1981.

"Challenge Interview: Walter Heller; Taxes and the State of the Economy." *Challenge* 12, no. 8 (May 1964): 20–24.

Citron, Rodger D. "Charles Reich's Journey from the *Yale Law Journal* to the *New York Times* Best-Seller List: The Personal History of *The Greening of America*." *New York Law Review* 52, no. 3 (2007–8): 387–416.

Clements, Kendrick. *The Life of Herbert Hoover*. New York: Palgrave MacMillan, 2010.

Cohen, Lizabeth. *A Consumer's Republic*. New York: Vintage Books, 2003.

Cole, H. L., and L. E. Ohanian. "New Deal Politics and the Persistence of the Great Depression: A General Equilibrium Analysis." *Journal of Political Economy* 112, no. 4 (August 2004): 779–816.

"Committee Findings." In *Recent Social Trends*, XI–LXXV.

Conway, Flo, and Jim Siegelman. *Dark Hero of the Information Age: In Search of Norbert Wiener, the Father of Cybernetics*. New York: Basic Books, 2005.

Cormier, Frank, and Walter J. Eaton. *Reuther.* Englewood Cliffs, NJ: Prentice Hall, 1970.

Council of Economic Advisers. *Economic Report of the President*. Washington, DC: US Government Printing Office, 1946.

Council of Economic Advisers. *Economic Report of the President*. Washington, DC: US Government Printing Office, 1947.

Council of Economic Advisers. *Economic Report of the President*. Washington, DC: US Government Printing Office, 1954.

Council of Economic Advisers. *Economic Report of the President*. Washington, DC: US Government Printing Office, 1955.

Council of Economic Advisers. *Economic Report of the President*. Washington, DC: US Government Printing Office, 1956.

Cowie, Jefferson. *The Great Exception: The New Deal and the Limits of American Politics*. Princeton, NJ: Princeton University Press, 2016.

Cowie, Jefferson. *Stayin' Alive: The 1970s and the Last Days of the Working Class*. New York: New Press, 2010.

Cowie, Jefferson, and Nick Salvatore. "The Long Exception: Rethinking the Place of the New Deal in American History." *International Labor and Working Class History*, no. 74 (2008): 3–32.

Critchlow, Donald T. *The Conservative Ascendancy: How the Republican Right Rose to Ascendancy in Modern America*. Lawrence: University Press of Kansas, 2011.

Cross, Wilbur. *John Diebold: Breaking the Confines of the Possible*. New York: James H. Heineman, 1965.

DaFoe, Allan. "On Technological Determinism: A Typology, Scope Conditions, and a Mechanism." *Science, Technology, and Human Values* 40, no. 6 (November 1915): 1047–76.

Dallek, Robert. *Flawed Giant: Lyndon Johnson and His Times*. New York: Oxford University Press, 1998.

Dallek, Robert. *Franklin D. Roosevelt: A Political Life*. New York: Viking, 2017.

Dallek, Robert. *Lyndon B. Johnson: Portrait of a President*. New York: Oxford University Press, 2004.

Dallek, Robert. *An Unfinished Life: John Kennedy, 1917–1963*. Boston: Little, Brown, 2003.

Davis, James L. *The Iron Puddler: My Life in the Rolling Mills*. Indianapolis: Bobbs-Merrill, 1922.

De Long, Bradford. "Keynesianism, Pennsylvania Avenue Style: Some Economic Consequences of the Employment Act of 1946." *Journal of Economic Perspectives* 10, no. 3 (Summer 1996): 41–53.

Dean, Alan L. "Ad Hoc Commissions for Policy Formulation." In *The Presidential Advisory System*, edited by Thomas E. Cronin and Sanford D. Greenberg, 101–21. New York: Harper and Row, 1969.

Dechter, Aimee R., and Glen Elder. "World War II Mobilization in Men's Work Lives: Continuity or Disruption for the Middle Class." *American Journal of Sociology* 110, no. 3 (November 2004): 761–64.

Denning, Peter J. "The Science of Computing: The ARPANET after Twenty Years." *American Scientist* 77, no. 6 (1989): 530, 532–34.

Diamond, Arthur M. "Edwin Mansfield's Contributions to the Economics of Technology." *Research Policy* 32, no. 9 (2003): 1607–17.

Dickson, Paul, and Thomas B. Allen. *The Bonus Army: An American Epic*. New York: Walker and Company, 2004.

Diebold, John. *Automation: The Advent of the Automatic Factory*. New York: D. Von Nostrand, 1952.

Diebold, John. "Automation—the New Technology." *Harvard Business Review*, November 1953, 63–71.

Diebold, John. *Beyond Automation: Managerial Problems of an Exploding Technology*. New York: Praeger, 1970.

Dimitri, Carolyn, Anne Effland, and Neilson Conklin. *The 20th Century Transformation of U.S. Agriculture and Farm Policy*. Washington, DC: US Department of Agriculture, June 2005. https://ageconsearch.umn.edu/bitstream/59390/2/eib3.pdf.

Directory of Labor Unions in the United States, Bulletin 901. Washington, DC: US Government Printing Office, 1947.

Dobney, Frederick J. "The Evolution of a Reconversion Policy: World War II and Surplus War Property Disposal." *Historian* 36, no. 3 (May 1974): 498–519.

Douglas, Paul H. *In the Fullness of Time: The Memoirs of Paul H. Douglas*. New York: Harcourt Brace Jovanovich, 1971.

"Drafting Problems and the Regulation of Featherbeds: An Imagined Dilemma." *Yale Law Journal* 73, no. 5 (April 1964): 812–49.

Dray, Philip. *There Is Power in a Union: The Epic Story of Labor in America*. New York: Doubleday, 2010.

Drucker, Peter. Foreword to *Beyond Automation*, by John Diebold, vii–ix. New York: Praeger, 1970.

Druckman, Mason. *Wayne Morse: A Political Biography*. Portland: Oregon Historical Society, 1997.

England, J. Merton. *A Patron for Pure Science*. Washington, DC: National Science Foundation, 1982.

Feurer, Rosemary, and Chad Pearson. "Introduction: Against Labor." In *Against Labor: How U.S. Employers Organized to Defeat Union Activism*, edited by Rosemary

Feurer and Chad Pearson, loc. 73–645 of 6969. Urbana: University of Illinois Press, 2017. Kindle.

Field, Alexander J. *A Great Leap Forward: 1930s Depression and U.S. Economic Growth.* New Haven, CT: Yale University Press, 2011.

Field, Alexander J. "The Impact of the Second World War on U.S. Productivity Growth." *Economic History Review* 61, no. 3 (August 2008): 672–94.

Field, Alexander J. "Technological Change and U.S. Productivity Growth in the Interwar Years." *Journal of Economic History* 66, no. 1 (2006): 203–36.

Finan, Christopher M. *Alfred E. Smith: The Happy Warrior.* New York: Hill and Wang, 2002.

Fine, Sidney. *Violence in the Model City: The Cavanagh Administration, Race Relations, and the Detroit Riot of 1967.* Lansing: Michigan State University Press, 2007.

Fishman, Betty C., and Leo Fishman. *Employment, Unemployment and Economic Growth.* New York: Thomas Y. Crowell, 1934.

Fleck, Robert F. "Democratic Opposition to the Fair Labor Standards Act of 1938." *Journal of Economic History* 62, no. 1 (March 2002): 25–54.

Ford, Martin. *The Rise of the Robots: Technology and the Threat of a Jobless Future.* New York: Basic Books, 2015.

Forstater, Mathew. "From Civil Rights to Economic Security: Bayard Rustin and the African American Struggle for Full Employment, 1945–1978." *International Journal of Political Economy* 36, no. 3 (Fall 2007): 63–74.

Frey, Carl Benedikt. *The Technology Trap: Capital, Labor, and Power in the Age of Automation.* Princeton, NJ: Princeton University Press, 2019.

Gaddis, Vincent Ray. "Herbert Hoover: Unemployment and the Public Sphere." PhD diss., Northern Illinois University, 2001.

Gay, Edwin F., and Leo Wolman. "Trends in Economic Organization." In *Recent Social Trends*, 218–67.

Gill, Corrington. *Unemployment and Technological Change.* Philadelphia: Works Progress Administration, National Research Project, April 1940.

Godin, Benoit. "Innovation without the Word: William F. Ogburn's Contribution to the Study of Technological Innovation." *Minerva* 48, no. 3 (September 2010): 277–307.

Goldberg, Stanley. "Inventing a Climate of Opinion: Vannevar Bush and the Decision to Build the Bomb." *Isis* 83, no. 3 (1992): 429–34.

Goldfield, David. "Writing the Sunbelt." *OAH Magazine of History* 18, no. 1 (October 2003): 5–10.

Goodwin, Doris Kearns. *No Ordinary Time: Franklin and Eleanor Roosevelt; The Home Front in World War II.* New York: Simon and Schuster, 1994. Kindle.

Gordon, Robert J. *The Rise and Fall of American Growth: The U.S. Standard of Living since the Civil War.* Princeton, NJ: Princeton University Press, 2016.

Graham, Hugh Davis. "The Ambiguous Legacy of American Presidential Commissions." *Public Historian* 7, no. 2 (Spring 1985): 5–25.

Grant, James. *Bernard Baruch: The Adventures of a Wall Street Legend.* New York: George Wiley and Sons, 1997.

Grant, James. *The Forgotten Depression: 1921; The Crash That Cured Itself.* New York: Simon and Schuster, 2015.

Greenberg, Cheryl. "Twentieth Century Liberalisms." In *Perspectives on Modern America*, edited by Harvard Sitkoff, 55–79. New York: Oxford University Press, 2001.

Haber, William. "The Strategy of Reconversion." *American Political Science Review* 38, no. 6 (1944): 1114–24.

Haber, William, and J. J. Joseph. "Unemployment Compensation." *Annals of the American Academy of Political and Social Science* 202 (March 1939): 22–35.

Halberstam, David. *The Best and the Brightest*. New York: Ballantine Books, 1969.

Hansen, Alvin H. *After the War—Full Employment*. National Resources Planning Board. Washington, DC: US Government Printing Office, 1942.

Harrington, Michael. *The Other America: Poverty in the United States*. New York: Simon and Schuster, 1962.

Hart-Landsberg, Martin. "Popular Mobilization and Progressive Policy Making: Lessons from World War II Price Control Struggles in the United States." *Science and Society* 67, no. 4 (Winter 2003/4): 399–428.

Heady, Ferrel. "The Reports of the Hoover Commission." *Review of Politics* 11, no. 3 (July 1949): 355–56, 375–78.

Heath, Jim F. "American War Mobilization and the Use of Small Manufacturers, 1939–1943." *Business History Review* 46, no. 3 (Autumn 1972): 295–319.

Heller, Walter H. *New Dimensions of Political Economy*. New York: W. W. Norton, 1966.

Hoeveler, J. David. *John Bascom and the Origins of the Wisconsin Idea*. Madison: University of Wisconsin Press, 2016. Kindle.

Hounshell, David. *From the American System to Mass Production, 1800–1932: The Development of Mass Production Technology in the United States*. Baltimore: Johns Hopkins University Press, 1984.

Humphrey, Hubert H. *The Education of a Public Man: My Life and Politics*. Minneapolis: University of Minnesota Press, 1991.

"Humphrey-Hawkins Full Employment Bill." In *CQ Almanac 1978*, 272–79. Washington, DC: Congressional Quarterly, 1979.

Hunnicutt, Benjamin Kline. "Kellogg's Six-Hour Day: A Capitalist Vision of Liberation through Managed Work Reduction." *Business History Review* 66, no. 3 (Autumn 1992): 475–522.

Hurlin, Ralph G., and Meredith Givens. "Shifting Occupational Patterns." In *Recent Social Trends*, 268–324.

Huthmacher, J. Joseph. *Senator Robert Wagner and the Rise of Urban Liberalism*. New York: Atheneum, 1968.

"In Memoriam: Leo Wolman." *Political Science Quarterly* 77, no. 3 (September 1962): 480–83.

Interbureau Committee and the Bureau of Agricultural Economics of the US Department of Agriculture. "Revolution on the Farm 1940." In *Readings in Technology and American Life*, edited by Carroll W. Pursell Jr., 372–79. New York: Oxford University Press, 1969.

Jacobs, Meg. "The Uncertain Future of American Politics, 1940 to 1973." In *American History*, edited by Eric Foner and Lisa McGerr, 151–74. Philadelphia: Temple University Press, 2011.

Javits, Jacob K. *Javits: The Autobiography of a Public Man*. With Rafael Steinberg. Boston: Houghton Mifflin, 1981.

Jeansonne, Glen. *Herbert Hoover: A Life*. New York: New American Library, 2016.

Jerome, Harry. *Mechanization in Industry*. Washington, DC: National Bureau of Economic Research, 1934.

John Lesinski Memorial Services. 82nd Congress, 2nd sess. Washington, DC: US Government Printing Office, 1951.

Jones, Matthew. "Freedom from Want." In *The Four Freedoms: Franklin D. Roosevelt and the Evolution of an American Idea*, edited by Jeffrey A. Engel, 125–64. New York: Oxford University Press, 2016.

Kaplan, A. D. H. *The Guarantee of Annual Wages*. Washington, DC: Brookings Institution, 1947.

Kashatus, William C. *Dapper Dan Flood: The Controversial Life of a Washington Power Broker*. College Station: Pennsylvania State University Press, 2010.

Kates, James. "Editor, Publisher, Citizen Soldier." *Journalism History* 44, no. 2 (Summer 2018): 79–89.

Katz, Michael D. *In the Shadows of the Poorhouse: A Social History of Welfare in America*. New York: Basic Books, 1996.

Katznelson, Ira. *Fear Itself: The New Deal and the Origins of Our Time*. New York: Liveright, 2013.

Katznelson, Ira, Kim Greiger, and Danile Kryder. "Limiting Liberalism: The Southern Veto in Congress, 1933–1950." *Political Science Quarterly* 108, no. 2 (Summer 1993): 283–306.

Kaufman, Bruce E. "Wage Theory, New Deal Labor Policy, and the Great Depression." *International Labor Review* 65, no. 3 (2012): 501–32.

Kennedy, David M. *Freedom from Fear: The American People in Depression and War, 1929–1945*. New York: Oxford University Press, 1999.

Key, V. O. "The Reconversion Phase of Demobilization." *American Political Science Review* 38, no. 6 (1944): 1137–53.

Keyssar, Alexander. "Unemployment before and after the Great Depression." *Social Research* 54, no. 2 (1987): 201–21.

King, Martin Luther, Jr. *"All Labor Has Dignity."* Edited and with an introduction by Michael K. Honey. Boston: Beacon, 2011.

Knapp, Joseph G. *Edwin G. Nourse: Economist for the People*. Danville, IL: Interstate Printers and Publishers, 1979.

Koistinen, Paul A. C. "Mobilizing the World War II Economy: Labor and the Industrial-Military Alliance." *Pacific Historical Review* 42, no. 4 (November 1973): 443–78.

Kuhn, David Paul. *Hard Hat Riot*. New York: Oxford University Press, 2020.

Lambert, Josiah Bartlett. *"If the Workers Took a Notion": The Right to Strike and American Political Development*. Ithaca, NY: Cornell University Press, 2005.

Landes, David S. *The Unbound Prometheus: Technological Change and Industrial Development in Western Europe from 1750 to the Present*. Cambridge: Cambridge University Press, 1969.

Latimer, Murray L. *Guaranteed Annual Wages Report for the President's Office of War Mobilization*. Washington, DC: US Government Printing Office, 1947.

Leuchtenberg, William E. *Franklin D. Roosevelt and the New Deal: 1932 to 1940*. New York: Harper Perennial, 1963.

Leuchtenburg, William E. *Herbert Hoover: The 31st President, 1929–1933*. The American Presidents Series. New York: Henry Holt, 2009.

Leuchtenburg, William E. *The White House Looks South: Franklin D. Roosevelt, Harry S. Truman, Lyndon B. Johnson.* Walter Lynwood Fleming Lectures in Southern History. Baton Rouge: Louisiana State University Press, 2005.

Lichtenstein, Nelson. "Class Politics and the State during World War Two." *International Labor and Working Class History* 58 (Fall 2000): 261–74.

Lichtenstein, Nelson. *A Contest of Ideas: Capital, Politics, and Labor.* Urbana: University of Illinois Press, 2013.

Lichtenstein, Nelson. *Labor's War at Home: The CIO in World War II.* Labor in Crisis Series. Philadelphia: Temple University Press, 2003.

Lichtenstein, Nelson. "Pluralism, Postwar Intellectuals, and the Demise of the Union Idea." In *The Great Society and the High Tide of Liberalism,* edited by Sidney Milkis and Jerome Mileur, 83–114. Amherst: University of Massachusetts Press, 2005.

Lichtenstein, Nelson. *State of the Union: A Century of America Labor.* Princeton, NJ: Princeton University Press, 2002.

Lichtenstein, Nelson. *Walter Reuther: The Most Dangerous Man in Detroit.* Urbana: University of Illinois Press, 1997.

Lippmann, Walter. "Discovered in Our Time." In *The Essential Lippmann: A Political Philosophy for Liberal Democracy,* edited by Clinton Rossiter and James Laree, 350–51. New York: Vintage Books, 1963.

Livermore, Charles H. "James G. Patton: Nineteenth-Century Populist, Twentieth-Century Organizer, Twenty-First-Century Visionary." PhD diss., University of Denver, 1976.

Lorant, John H. "Technological Change in American Manufacturing during the 1920's." *Journal of Economic History* 27, no. 2 (1967): 243–46.

Malone, Michael S. *The Intel Trinity: How Robert Noyce, Gordon Moore, and Andy Grove Built the World's Most Important Company.* New York: Harper Business, 2014.

Mansfield, Edwin. "Size of the Firm, Market Structure, and Innovation." *Journal of Political Economy* 71, no. 6 (December 1963): 556–76.

Mansfield, Edwin. "Technical Change and the Rate of Imitation." *Econometrica* 29, no. 4 (October 1961): 741–66.

Mayhew, David R. "The Long 1950s as a Policy Era." In *The Politics of Major Policy Reform in Postwar America,* edited by Jeffrey A. Jenkins and Sidney M. Milkis, 27–47. New York: Cambridge University Press, 2014.

McElvaine, Robert S. *The Great Depression: America, 1929–1941.* New York: Three Rivers, 1984.

McGerr, Michael. *A Fierce Discontent: The Rise and Fall of the Progressive Movement in America, 1870–1920.* New York: Oxford University Press, 2003.

McGirr, Lisa. *Suburban Warriors: The Origins of the New American Right.* Princeton, NJ: Princeton University Press, 2001.

Meyer, Steve. "An Economic Frankenstein: UAW Workers' Responses to Automation at the Ford Brook Park Plant in the 1950s." *Michigan Historical Review* 28, no. 1 (2002): 63–89.

Mileur, Jerome M. "The Great Society and the Demise of New Deal Liberalism." In *The Great Society and the High Tide of Liberalism,* edited by Sidney M. Milkis and Jerome M. Mileur, 411–56. Amherst: University of Massachusetts Press, 2005.

Milkis, Sidney M. "Franklin Roosevelt, the Economic Constitutional Order and the New Politics of Presidential Leadership." In *The New Deal and the Triumph of Liberalism*, edited by Sidney M. Milkis and Jerome M. Mileur, 31–72. Amherst: University of Massachusetts Press, 2002.

Milkis, Sidney M., and Jerome M. Mileur, eds. *The Great Society and the High Tide of Liberalism*. Amherst: University of Massachusetts Press, 2005.

Milkis, Sidney M., and Jerome M. Mileur, eds. *The New Deal and the Triumph of Liberalism*. Amherst: University of Massachusetts Press, 2002.

Miller, Carolyn. "Learning from History: World War II and the Culture of High Technology." *Journal of Business and Technical Communication* 12, no. 3 (July 1998): 288–95.

Milton, David. *The Politics of U.S. Labor: From the Great Depression to the New Deal*. New York: Monthly Review Press, 1982.

Misa, Thomas J. *A Nation of Steel: The Making of Modern America*. Baltimore: Johns Hopkins University Press, 1999.

Molnar, F. "The 1974–75 Recession in the USA: A Lot of Facts and Some Lessons." *Acta Oeconomica* 17, no. 2 (1976): 177–201.

National Commission on Technology, Automation, and Economic Progress. *The Employment Impact of Technological Change: Appendix II to "Technology and the American Economy."* Washington, DC: US Government Printing Office, February 1966.

National Commission on Technology, Automation, and Economic Progress. *Statements Relating to the Impact of Technological Change: Appendix VI to "Technology and the American Economy."* Washington, DC: US Government Printing Office, 1966.

National Commission on Technology, Automation, and Economic Progress. *Technology and the American Economy*. Washington, DC: US Government Printing Office, 1966.

Nelson, Daniel. "How the UAW Grew." *Labor History* 35 (1994): 5–24.

Newman, Roger K. *Hugo Black: A Biography*. New York: Pantheon Books, 1994.

Newton, Jim. *Eisenhower: The White House Years*. New York: Doubleday, 2011. Kindle.

Noble, David F. *Forces of Production: A Social History of Industrial Automation*. New York: Alfred A. Knopf, 1984.

Noble, David F. *Progress without People*. Toronto: Between the Lines, 1995. Kindle.

Nuclear Power Economics—Analysis and Comment—1964. Joint Committee on Atomic Energy. Washington, DC: US Government Printing Office, October 1964.

Nye, David E. *America's Assembly Line*. Cambridge, MA: MIT Press, 2013. Kindle.

O'Donnell, Edward T. *Henry George and the Crisis of Inequality: Progress and Poverty in the Gilded Age*. New York: Columbia University Press, 2015.

Ogburn, W. F. "The Influence of Invention and Discovery." In *Recent Social Trends*, 122–66.

Ogburn, William F. *Social Change with Respect to Culture and Original Nature*. New York: B. W. Huebsch, 1922.

Ohl, John Kennedy. *Hugh Johnson and the New Deal*. DeKalb: Northern Illinois University Press, 1985.

Oliphant, Thomas, and Curtis Wilkie. *The Road to Camelot: Inside JFK's Five-Year Campaign*. New York: Simon and Schuster, 2017.

O'Neill, William L. *Coming Apart: An Informal History of American in the 1960's.* Chicago: Ivan Dee, 1971.

Ortiz, Stephen R, ed. *Veterans' Policies, Veterans' Politics: New Perspectives on Veterans in the United States.* Tallahassee: University of Florida Press, 2012.

Owens, Larry. "The Counterproductive Management of Science in the Second World War: Vannevar Bush and the Office of Scientific Development." *Business History Review* 68, no. 4 (Winter 1994): 515–19.

Palmer, Niall. *The Twenties in America: Politics and History.* Edinburgh: Edinburgh University Press, 2006.

Patterson, James T. *The New Deal and States: Federalism in Transition.* Princeton, NJ: Princeton University Press, 1969.

Perlstein, Rick. *Before the Storm: Barry Goldwater and the Unmaking of the American Consensus.* New York: Nation Books, 2001.

Perlstein, Rick. *Nixonland: The Rise of a President and the Fracturing of America.* New York: Scribner, 2008.

Perry, George L., Edward F. Denison, and Robert M. Solow. "Labor Force Structure, Potential Output, and Productivity." *Brookings Papers on Economic Activity,* no. 3 (1971): 533–78.

Phillips-Fein, Kim. "Conservatism: A State of the Field." *Journal of American History* 98, no. 3 (December 2011): 723–43.

Phillips-Fein, Kim. *Invisible Hands: The Businessman's Crusade against the New Deal.* New York: W. W. Norton, 2009.

Phipps, Charles. "The Early History of ICs at Texas Instruments: A Personal View." *IEEE Annals of the History of Computing* 34, no. 1 (January 2012): 37–47.

Polsby, Nelson W. *Political Innovation in America: The Politics of Policy Initiation.* New Haven, CT: Yale University Press, 1984.

Popper, Frank. *The President's Commissions.* New York: Twentieth Century Fund, 1970.

Postel, Charles. *The Populist Vision.* New York: Oxford University Press, 2015.

Price, Daniel N. "Unemployment Insurance, Then and Now, 1935–1985." *Social Security Bulletin* 48, no. 10 (October 1985): 22–32.

"Progress in American Labor Legislation." In *Proceedings of the 10th Annual Convention of the Association of Governmental Labor Officials of the United States and Canada,* 48–57. Washington, DC: US Government Printing Office, 1923.

Prout, Jerry. *Coxey's Crusade for Jobs: Unemployment in the Gilded Age.* DeKalb: Northern Illinois University Press, 2016.

Public Papers of the President of the United States: Lyndon B. Johnson; 1966. Book 1. Washington, DC: US Government Printing Office, 1967.

Pursell, Carroll. *Technology in Postwar America: A History.* New York: Columbia University Press, 2007.

Pursell, Carroll W., Jr. *Readings in Technology and American Life.* New York: Oxford University Press, 1969.

Quigley, William P. *Ending Poverty as We Know It.* Philadelphia: Temple University Press, 2008.

Raskin, A. H. "The Great Society." *Vital Speeches* 32, no. 18 (July 1, 1966): 554–59.

Ratner, Mark A. "Especially about the Future." In *The Fabulous Future: America and the World in 2040,* edited by Gary Saul Morson and Morton Schapiro, 99–112. Evanston, IL: Northwestern University Press, 2015.

Rauchway, Eric. *Winter War: Hoover, Roosevelt, and the First Clash Over the New Deal.* New York: Basic Books, 2018.

Recent Social Trends in the United States. Report of the President's Research Committee on Social Trends. Vol. 1. New York: McGraw Hill, 1933.

Reich, Leonard S. "From the Spirit of St. Louis to the SST: Charles Lindbergh, Technology and the Environment." *Technology and Culture* 36, no. 2 (1995): 351–93.

Remini, Robert V. *The House: The History of the House of Representatives.* New York: HarperCollins, 2006.

"Report of the Advisory Committee on Employment Statistics." *Bulletin of the United States Bureau of Labor Statistics,* no. 542 (1931): i–vi, 1–31.

Report of the President's Conference on Unemployment. Washington, DC: US Government Printing Office, 1921.

Reuther, Walter. "Policies for Automation: A Labor Viewpoint." *Annals of the American Academy of Political and Social Science* 340 (March 1962): 100–109.

Rodgers, Daniel T. *Atlantic Crossings: Social Politics in a Progressive Age.* Cambridge, MA: Belknap Press of Harvard University Press, 1998.

Roediger, David R., and Philip S. Foner. *Our Own Time: A History of American Labor and the Working Day.* New York: Verso, 1989.

Rose, Mark H. "The Historiography of Technology and Public Policy." *Public Historian* 10, no. 2 (1988): 27–47.

Rosen, Elliot A. *Roosevelt, the Great Depression, and the Economics of Recovery.* Charlottesville: University of Virginia Press, 2005. Kindle.

Ruebhausen, Oscar M., and Robert B. Von Mehren. "The Atomic Energy Act and the Production of Atomic Power." *Harvard Law Review* 66, no. 8 (1953): 1450–96.

Salvatore, Nick. *Eugene V. Debs: Citizen and Socialist.* Urbana: University of Illinois Press, 1982.

Samuel, Howard D. "Troubled Passage: The Labor Movement and the Fair Labor Standards Act." *Labor Policy Review* 62, no. 1 (March 2002): 25–54.

Sautter, Udo. "Government and Unemployment: The Use of Public Works before the New Deal." *Journal of American History* 73, no. 1 (1986): 59–86.

Sautter, Udo. *Three Cheers for the Unemployed: Government and Unemployment before the New Deal.* New York: Cambridge University Press, 1995.

Schlesinger, Arthur M., Jr. *A Thousand Days: John F. Kennedy in the White House.* Boston: Houghton Mifflin, 1965.

Segal, Howard P. *Technological Utopianism in American Culture.* Syracuse, NY: Syracuse University Press, 2005.

Senate Committee on Education and Labor. *Causes of Unemployment.* Report No. 2072. 70th Congress, 2nd sess. February 25, 1929.

Senate Committee on Post-war Economic Policy and Planning. *Report on Postwar Economic Policy and Planning Pursuant to S. Res. 102.* S. Report 539, pt. 2. 78th Congress, 2nd sess. February 9, 1944.

Senate Special Committee on Unemployment. *Report of the Special Committee on Unemployment Problems.* Washington, DC: US Government Printing Office, March 23, 1960.

Shaffer, Helen B. "Guaranteed Annual Wage." *Editorial Research Reports* 1 (January 21, 1953): 43–59.

Shales, Amity. *Great Society: A New History*. New York: HarperCollins, 2019.

Shaw, John T. *JFK in the Senate: Pathway to the Presidency*. New York: St. Martin's, 2013.

Shermer, Elizabeth Tandy. "Counter-organizing the Sunbelt: Right-to-Work Campaigns and Anti-union Conservatism, 1943–1958." *Pacific Historical Review* 78, no. 1 (February 2009): 81–118.

Shermer, Elizabeth Tandy. *Sunbelt Capitalism: Phoenix and the Transformation of American Politics*. Philadelphia: University of Pennsylvania Press, 2013.

Slesinger, Reuben E. "The Pace of Automation: An American View." *Journal of Industrial Economics* 6, no. 3 (1958): 241–61.

Slichter, Sumner H. "The Current Labor Policies of American Industries." *Quarterly Journal of Economics* 3, no. 3 (May 1929): 393–435.

Smiley, Gene. *Rethinking the Depression*. Chicago: Ivan Dee, 2002.

Smith, Jason Scott. *Building New Deal Liberalism: The Political Economy of Public Works, 1933–1956*. Cambridge: Cambridge University Press, 2006.

Solberg, Carl. *Hubert Humphrey: A Biography*. St. Paul, MN: Borealis Books, 1984.

Solow, Robert M. "A Contribution to the Theory of Economic Growth." *Quarterly Journal of Economics* 70, no. 1 (February 1956): 65–94.

Solow, Robert M. "Technical Change and the Aggregate Production Function." *Review of Economics and Statistics* 39, no. 3 (1957): 312–20.

Sorensen, Theodore. *Kennedy*. New York: Harper Perennial, 1965.

Statistical Abstract of the United States, 1944–45. Washington, DC: US Government Printing Office, 1945.

Steigerwald, David. "Walter Reuther, the UAW, and the Dilemmas of Automation." *Labor History* 51, no. 3 (2010): 429–53.

Stein, Judith. *Pivotal Decade: How the United States Traded Factories for Finance in the Seventies*. New Haven, CT: Yale University Press, 2010. Kindle.

Stein, Leon. *The Triangle Fire*. Ithaca, NY: Cornell University Press, 1962.

Stershner, Bernard. "Victims of the Great Depression: Self-Blame/Non-self-blame, Radicalism, and Pre-1929 Experiences." *Social Science History* 1, no. 2 (Winter 1977): 137–77.

Stewart, Ethelbert. "Wastage of Men." *Monthly Labor Review* 19, no. 1 (1924): 1–8.

Stone, Ralph A., "Two Illinois Senators among the Irreconcilables." *Mississippi Valley Historical Review* 50, no. 3 (1963): 443–47.

Stricker, Frank. *American Unemployment: Past, Present, and Future*. Urbana: University of Illinois Press, 2020.

Technology in Our Economy: Investigation of Concentration of Economic Power. Monograph No. 22 of the Temporary National Economic Committee. Washington, DC: US Government Printing Office, 1941.

Terborgh, George. *The Automation Hysteria*. New York: W. W. Norton, 1965.

Thomas, Norman C., and Harold L. Wolman. "Policy Formulation in the Institutionalized Presidency: The Johnson Task Forces." In *The Presidential Advisory System*, edited by Thomas E. Cronin and Sanford D. Greenberg, 107–24. New York: Harper and Row, 1969.

Thompson, Warren S., and P. K. Whelpton. "The Population of the Nation." In *Recent Social Trends*, 1–58.

Townsend, Edwin T. "The Human Equation." *Challenge* 9, no. 5 (February 1961): 16–20.

Trubek, Anne, ed. *Voices from the Rust Belt*. New York: Picador, 2018.

Tugwell, Rexford G. *Industry's Coming of Age*. New York: Harcourt, Brace, 1927.

Tugwell, Rexford G. "The Theory of Occupational Obsolescence." *Political Science Quarterly* 46, no. 2 (June 1931): 181–225.

US Bureau of Labor Statistics. *Technological Trends in Major American Industries*. Bulletin of the United States No. 1474. February 1966. https://fraser.stlouisfed.org/title/4935.

US Bureau of Labor Statistics. *Technological Trends in 36 Major American Industries*. Washington, DC: US Government Printing Office, June 1964.

US Bureau of Labor Statistics. *Work Stoppages Caused by Labor-Management Disputes in 1945*. Bulletin No. 878. May 20, 1946.

US Bureau of Labor Statistics. *Work Stoppages Caused by Labor-Management Disputes in 1946*. Bulletin No. 918. June 1947. https://fraser.stlouisfed.org/files/docs/publications/bls/bls_0918_1947.pdf.

US Congressional Budget Office. *The Productivity Slowdown: Causes and Response*. June 1, 1981.

US Department of Agriculture. *Technology on the Farm*. August 1940.

US Departments of Commerce and Labor, *Report of the President's Advisory Committee on Labor-Management Policy: The Benefits and Problems Incident to Automation and Other Technological Advances*. January 11, 1962.

US House of Representatives. *Communication from the President of the United States: Supplemental Estimate of Appropriations for the Office of War Mobilization and Reconversion*. Document No. 442. 79th Congress, 2nd sess. January 31, 1946.

US Senate. *Create a Temporary National Economic Committee*. S. Rept. No. 1991. 75th Congress, 3rd sess. June 6, 1938.

US Senate. *Report Pursuant to S. Res. 219, Causes of Unemployment*. S. Report No. 2072. 70th Congress, 2nd sess. February 25, 1929.

US Senate. *Report to Accompany S. J. Res. 300*. Report No. 1991. 75th Congress, 3rd sess. June 6, 1938.

Usselman, Steven W. "Research and Development in the United States since 1900: An Interpretive History." Paper presented at the Economic History Workshop, Yale University, New Haven, CT, November 11, 2013.

Vatter, Harold G. *The U.S. Economy in World War II*. New York: Columbia University Press, 1985.

Von Drehle, David. *Triangle: The Fire That Changed America*. New York: Grove, 2003.

Vought, Hans P. *The Bully Pulpit and the Melting Pot*. Macon, GA: Mercer University Press, 2004.

Walker, L. C. "The Share the Work Movement." *Annals of the Academy of Political and Social Science* 165 (January 1933): 13–19.

Wall, Wendy L. *Inventing the "American Way": The Politics of Consensus from the New Deal to the Civil Rights Movement*. New York: Oxford University Press, 2008.

Wang, Jessica. *American Science in an Age of Anxiety*. Chapel Hill: University of North Carolina Press, 1984.

Warren, Dorian T. "The Politics of Labor Policy Reform." In *The Politics of Major Policy Reform in Postwar America*, edited by Jeffrey A. Jenkins and Sidney M. Milkis, 103–13. New York: Cambridge University Press, 2014.

Wasem, Ruth Ellen. *Tackling Unemployment*. Kalamazoo, MI: W. E. Upjohn Institute for Employment Research, 2013. Kindle.

Waterhouse, Ben. *Lobbying America*. Princeton, NJ: Princeton University Press, 2014.

Waters, Malcolm. *Daniel Bell*. New York: Routledge, 1996.

Waud, Roger N. "Note: A Stagnationist View of the 1960–61 Recession." *American Economist* 6 (May 1962): 21–25.

Web, Derek A. "The Natural Rights of Liberalism of Franklin Delano Roosevelt." *American Journal of Legal History* 55, no. 3 (September 2015): 331–346.

Weber, John. *Homing Pigeons, Cheap Labor, and Frustrated Nativists*. Chapel Hill: University of North Carolina Press, 2015.

Weintraub, David. "Unemployment and Increasing Productivity." In *Technological Trends and National Policy*, 67–87. Philadelphia: US Government Printing Office, 1937.

Weintraub, David. *The Work and Publications of the WPA National Research Project on Reemployment Opportunities and Recent Changes in Industrial Technique*. Philadelphia: Works Progress Administration, 1941.

Weir, Margaret. *Politics and Jobs*. Princeton, NJ: Princeton University Press, 1992.

White, Theodore. *In Search of History*. New York: Harper and Row, 1978.

White, Theodore. *The Making of the President 1960*. New York: HarperCollins, 1961.

White, Theodore. *The Making of the President 1964*. New York: HarperCollins, 1965.

Whyte, Kenneth. *Hoover: An Extraordinary Life in Extraordinary Times*. New York: Alfred A. Knopf, 2017.

Wiener, Norbert. *Cybernetics*. Cambridge, MA: MIT Press, 1948.

Wilkerson, Isabel. *The Warmth of Other Suns: The Epic Story of America's Great Migration*. New York: Vintage Books, 2011.

Wilson, Gregory S. "Before the Great Society: Liberalism, Deindustrialization and Area Development in the United States, 1933–1965." PhD diss., Ohio State University, 2001.

Wilson, Gregory S. *Communities Left Behind: The Area Redevelopment Administration, 1945–1965*. Knoxville: University of Tennessee Press, 2009.

Woirol, Gregory R. "The Machine Taxers." *History of Political Economy* 50, no. 4 (December 2018): 709–33.

Woirol, Gregory R. *The Technological Unemployment and Structural Unemployment Debates*. Westport, CT: Greenwood, 1996.

Wolanin, Thomas R. *Presidential Advisory Commissions*. Madison: University of Wisconsin Press, 1975.

Wooten, James A. *The Employee Retirement Income Security Act of 1974: A Political History*. Berkeley: University of California Press, 2004.

Wright, Gavin. *Old South, New South: Revolutions in the Southern Economy since the Civil War*. Baton Rouge: Louisiana State University Press, 1986.

Young, Nancy Beck. "'Do Something for the Soldier Boys': Congress, the G.I. Bill of Rights, and the Contours of Liberalism." In *Veterans' Policies, Veterans' Politics: New Perspectives on Veterans in the Modern United States*, edited by Stephen R. Ortiz, 199–221. Tallahassee: University of Florida Press, 2012.

Zachary, G. Pascal. *Endless Frontier: Vannevar Bush, Engineer of the American Century.* New York: Free Press, 1997.

Zegart, Amy B. "Black Boxes: Toward a Better Understanding of Presidential Commissions." *Presidential Studies Quarterly* 34, no. 2 (June 2004): 366–93.

Zelizer, Julian E. *The Fierce Urgency of Now: Lyndon Johnson, Congress, and the Battle for the Great Society.* New York: Penguin, 2015.

Zieger, Robert H. *For Jobs and Freedom: Race and Labor in America since 1865.* Lexington: University Press of Kentucky, 2007.

Zieger, Robert H. *Republicans and Labor: 1919–1929.* Lexington: University Press of Kentucky, 1969.

Congressional Hearings

House Committee on Banking and Currency. *Legislation to Relieve Unemployment.* 85th Congress, 2nd sess., April 14, 15, 21, and 22, and May 20, 1958.

House Committee on Banking, Housing, and Urban Affairs. *Full Employment and Balanced Growth Act of 1976.* 94th Congress, 2nd sess., May 20, 1976.

House Committee on Expenditures in the Executive Departments. *Full Employment Act of 1945.* 79th Congress, 1st sess., September 25 and October 16, 25, 1945.

House Committee on the Budget. *Hearing on Machines, Artificial Intelligence and the Workforce: Recovering and Readying Our Economy for the Future.* 116th Congress, 2nd sess., September 10, 2020.

House Judiciary Committee. *Unemployment in the United States.* 71st Congress, 2nd sess., June 11–12, 1930.

House Committee on Labor. *Hearings on H. Res. 49.* 74th Congress, 2nd sess., February 1, 14, 17, 20, and March 2, 1936.

House Committee on Labor. *Report to Accompany H. Res. 49.* 74th Congress, 2nd sess., no. 2685, May 19, 1936.

House Select Subcommittee on Labor, Committee on Education and Labor. *Hearings on H.R. 10310 and Related Bills.* 88th Congress, 2nd sess., April 14, 1964.

House Special Committee on Postwar Economic Policy and Planning. *Postwar Economic Policy and Planning.* 79th Congress, 1st sess., March 13, 1945.

House Subcommittee of the Committee on Ways and Means. *Unemployment Insurance, H.R. 7659.* 73rd Congress, 2nd sess., March 21, 1934.

House Ways and Means Committee. *Economic Security Act, H.R. 4120.* 74th Congress, 1st sess., January 21, 1935.

Joint Committee on the Economic Report. *Automation and Technological Change.* 84th Congress, 2nd sess., January 5, 1956.

Joint Committee on the Economic Report. *Hearings on Automation and Technological Change.* 84th Congress, 1st sess., October 14, 17, 1955.

Joint Committee on the Economic Report. *January 1954 Economic Report of the President.* 83rd Congress, 2nd sess., February 17, 1954.

Joint Committee on the Economic Report. *Report of the Joint Committee on the Economic Report on the January 1955 Economic Report of the President.* 84th Congress, 1st sess., March 14, 1955.

Joint Special Committee on Postwar Economic Policy and Planning. *Postwar Economic Policy and Planning.* 78th Congress, 2nd sess., March 13, June 16, July 27, 1944.

Senate Committee on Education and Labor. *Unemployment in the United States.* 70th Congress, 2nd sess. February 7, 1929

Senate Committee on Labor. *Investigation of Unemployment Caused by Labor-Saving Devices in Industry.* 74th Congress, 2nd sess., February 17, 1936.

Senate Committee on Labor and Public Works. *Dr. Garth Mangum to be Executive Secretary of the National Commission on Technology, Automation, and Economic Progress.* 89th Congress, 1st sess., March 5, 1965.

Senate Committee on Military Affairs. *Mobilization and Demobilization Problems.* 78th Congress, 2nd sess., April 26, June 9, 1944.

Senate Committee on Military Affairs. *Technological Mobilization.* 75th Congress, 2nd sess., October 13, 1942.

Senate Labor and Public Welfare Subcommittee on Employment and Manpower. *Nation's Manpower Revolution—National Commission on Automation and Technological Progress.* 88th Congress, 2nd sess., July 6, 1964.

Senate Labor Subcommittee on Labor and Public Welfare. *Area Redevelopment Hearings on S. 2663.* 84th Congress, 2nd sess., January 4 and 23, 1956.

Senate Special Committee on Unemployment Problems. *Hearings on Unemployment Problems, Part 1.* 86th Congress, 1st sess., October 5, 1959.

Senate Special Committee on Unemployment Problems. *Hearings on Unemployment Problems, Part 6.* 86th Congress, 1st sess., November 16, 1959.

Senate Special Committee on Unemployment Problems. *Hearings on Unemployment Problems, Part 8.* 86th Congress, 1st sess., November 3, 1959.

Senate Special Committee to Investigate Unemployment and Relief. *Unemployment and Relief.* 75th Congress, 3rd sess., April 8, 1938.

Senate Subcommittee on Employment and Manpower, Committee on Labor and Public Welfare. *Nation's Manpower Revolution.* 88th Congress, 1st sess., May 20, 21, 22; June 6; September, 19, 20; October 28; November 13, 1963.

Senate Subcommittee on Full Employment, Senate Committee on Banking and Currency. *Full Employment of 1945.* 79th Congress, 1st sess., July 30, July 31, August 24, October 16, 1945.

Senate Temporary National Economic Committee. *Investigation of Concentration of Economic Power.* 75th Congress, 3rd sess., December 1–3, 1938; 76th Congress, 3rd sess., April 6, 26, 1940.

Congressional Record

Congressional Record. 67th Congress, 1st sess. Vol. 61, pts. 2, 4, 5.

Congressional Record. 67th Congress, 2nd sess., 1921. Vol. 61, pts. 2, 5.

Congressional Record. 70th Congress, 1st sess., 1928. Vol. 69, pts. 3, 4, 5, 6, 7.

Congressional Record. 71st Congress, 2nd sess., 1930. Vol. 72, pt. 5.

Congressional Record. 71st Congress, 3rd sess., 1931. Vol. 74, pt. 6.

Congressional Record. 72nd Congress, 2nd sess., 1933. Vol. 76, pts. 3, 4.

Congressional Record. 73rd Congress, 1st sess., 1933. Vol. 77, pts. 2,5.

Congressional Record. 73rd Congress, 2nd sess., 1934. Vol. 78, pt. 4.

Congressional Record. 74th Congress, 1st sess., 1935. Vol. 79, pts. 1, 3, 5.

Congressional Record. 74th Congress, 2nd sess., 1936. Vol. 80, pts. 3, 4, 10.

Congressional Record. 75th Congress, 1st sess., 1937. Vol. 81, pt. 7.

Congressional Record. 75th Congress, 1st sess., 1937. Vol. 82, pt. 2.

Congressional Record. 75th Congress, 3rd sess., 1938. Vol. 83, pt. 2.

Congressional Record. 77th Congress, 2nd sess., 1942. Vol. 88, pt. 6.

Congressional Record. 78th Congress, 2nd sess., 1944. Vol. 90, pts. 2, 3.

Congressional Record. 79th Congress, 1st sess., 1945. Vol. 91, pts. 7, 9.

Congressional Record. 79th Congress, 2nd sess., 1946. Vol. 92, pts. 1, 8.

Congressional Record. 80th Congress, 1st sess., 1947. Vol. 93, pt. 2.

Congressional Record. 81st Congress, 1st sess., 1949. Vol. 95., pt. 7.

Congressional Record. 83rd Congress, 1st sess., 1953. Vol. 99., pt. 4.

Congressional Record. 84th Congress, 1st sess., 1955. Vol. 101, pts. 4, 9.

Congressional Record. 84th Congress, 2nd sess., 1956. Vol. 102, pts. 6, 7.

Congressional Record. 86th Congress, 2nd sess., 1960. Vol. 106, pt. 6.

Congressional Record. 88th Congress, 1st sess., 1963. Vol. 109, pts. 5, 9, 10, 16, 17.

Congressional Record. 88th Congress, 2nd sess., 1964. Vol. 110, pts. 1, 2, 4, 12, 13, 14.

Congressional Record. 89th Congress, 2nd sess., 1966. Vol. 112, pt. 2.

Congressional Record. 90th Congress, 2nd sess., 1968. Vol. 114, pt. 18.

Congressional Record. 93rd Congress, 2nd sess., 1974. Vol. 120, pts. 16, 22.

INDEX

The letter *f* following a page number denotes a figure.

CPSIA information can be obtained
at www.ICGtesting.com
Printed in the USA
LVHW100556160622
721414LV00016B/343/J

9 781501 763991